AUTOMOTIVE
RUST REPAIR
& PREVENTION

By Dennis W. Parks

motorbooks

Dedication

To all of the hot rodders, automotive enthusiasts, and anyone else who has fought the fight against a rusty automobile.

Brimming with creative inspiration, how-to projects, and useful information to enrich your everyday life, Quarto Knows is a favorite destination for those pursuing their interests and passions. Visit our site and dig deeper with our books into your area of interest: Quarto Creates, Quarto Cooks, Quarto Homes, Quarto Lives, Quarto Drives, Quarto Explores, Quarto Gifts, or Quarto Kids.

Motorbooks titles are also available at discount for retail, wholesale, promotional, and bulk purchase. For details, contact the Special Sales Manager by email at specialsales@quarto.com or by mail at The Quarto Group, Attn: Special Sales Manager, 401 Second Avenue North, Suite 310, Minneapolis, MN 55401 USA.

10 9 8 7 6 5 4 3 2 1

ISBN: 978-0-7603-5899-3

Library of Congress Cataloging-in-Publication Data

Names: Parks, Dennis, 1959- author.
Title: Automotive rust repair and prevention / by Dennis Parks.
Description: Minneapolis, MN : Motorbooks, an imprint of The Quarto Group,
    2018. | Includes bibliographical references and index.
Identifiers: LCCN 2017059861 | ISBN 9780760358993 (sc)
Subjects: LCSH: Automobiles--Bodies--Maintenance and repair. |
    Automobiles--Corrosion. | Corrosion and anti-corrosives.
Classification: LCC TL255 .P3595 2018 | DDC 629.2/60288--dc23
LC record available at https://lccn.loc.gov/2017059861

Acquisitions Editor: Zack Miller
Project Manager: Jordan Wiklund
Design Manager: Brad Springer
Layout Designer: Danielle Smith-Boldt

Printed in China

Contents

Acknowledgments

A very sincere thank you goes out to everyone who helped with this book, as I could not have done it without your help. In no particular order, thanks again to Jim Miller, Duane Wissman, and the staff at Jerry's Auto Body; Keith Moritz and the staff at Morfab Customs; Dylan Stevens; and Kevin and Wendy Brinkley at Harding's Auto Body.

I also need to thank Roger Ward at Bad Paint Company and John Kimbrough for all of the information they have provided over the years and for the proofreading they have done on my manuscripts.

Dennis W. Parks

Introduction

Unless you have always lived in a climate that is free of humidity, rain, road salt, chipped, or scratched paint, or any of the other things that cause it, chances are that you have seen, driven, or perhaps even owned an automobile that was affected by rust. Although some locations are more susceptible to the formation of rust than others, rust can be found most anywhere.

Rust can be prevented, and ways to do that will be discussed in this book. However, the goal for most of this book is learn how to detect, stop, and repair rust before it gets totally out of hand. Ultimately, the quick answer for getting rid of rust is cutting it all out and welding in new sheetmetal body panels. If money is no object, this is the best method. To find out how to fix rust using that method, please refer to my book, *The Complete Guide to Auto Body Repair, 2nd Edition*. Multiple examples of removing and replacing entire panels are included in that book.

However, this book is geared toward the guys and gals who are simply trying to keep their daily driver alive and presentable. With that in mind, the repairs shown in this book will be made *without* the use of a welder or spray gun. You are not resurrecting a collectible classic, building a boulevard cruiser, or even a weekend car show participant. You are merely trying to spend the least amount of money necessary on your vehicle in order to keep it drivable and respectable looking before it checks in as a resident at the local salvage yard.

In Chapter 1, we'll discuss the science of rust—what causes it and where it is usually found. This will help you detect it early on, in hopes of minimizing the damage, and you can take action to prevent it from returning. In Chapter 2, required tools and materials will be the major topic. Knowing what you need, being familiar with their use, and having them on hand makes most any task easier. Uncovering rust (minor, moderate, and advanced) and repairing it will be covered in Chapters 3 and 4. Finally, Chapter 5 discusses how to prevent new or reoccurring rust issues.

Rust sure isn't pretty, but you don't have to live with it. With a little bit of effort and a few tools, you can make it go away.

I hope this book gets you motivated about doing your own auto rust repair and gives you the confidence to do it yourself. Thank you for buying a copy.

Chapter 1
What Is Rust?

Admittedly, this book is not really going to tell you everything you may (but probably don't want to) know about rust. Countless metallurgy books delve into the scientific metallic aspects of rust, much deeper than this book or its author care to. What this book will do is provide you with a basic understanding of what causes rust to appear on your vehicle and how to deal with it, keeping your vehicle looking its best in the process.

CAUSES

Iron or an iron alloy, when exposed to moisture and oxygen for a long enough time, will turn to iron oxide, otherwise known as ferrous oxide. To most of us, this reaction or the product of this reaction is commonly known as rust, but there are two types of rust: ferrous oxide, caused by moisture and oxygen, and iron hydroxide, caused when sulfur dioxide and/or carbon dioxide combine with water that comes in contact with iron or iron alloys.

Now what makes this somewhat confusing is that pure water or dry oxygen do not have this common effect on iron. However, most, but not all, of the earth's surface receives some amount of humidity. It is very low in some areas and very high in others, but where humidity is present, the air is no longer "dry" oxygen. Likewise, "pure" water is probably not something that our vehicles come in contact with very often.

So, exposure to moisture and oxygen will cause iron to turn into iron oxide. Other elements can and usually will accelerate the rusting process. A common one that

This late 1940s Studebaker pickup truck looks pretty rusty, and therefore may not appeal to everyone, even if you do like old pickup trucks. However, the body is actually pretty solid as the rust is merely surface rust. As long as the sheet metal is reasonably solid, the truck can be cleaned up and resurrected.

accelerates deterioration of iron is the presence of salt. Remember this when you move to a saltwater beach area or when you believe that the highway department needs to do more to prepare the roads for that next big snowstorm.

Trapped Moisture

Whenever moisture gets trapped on a piece of bare metal or between multiple pieces or layers, it will typically lead to the formation of rust, unless it is removed quickly. This is typically the reason a vehicle's chassis or frame gets rusty. In several places on a vehicle, structural metal components are welded or bolted to other similar components (spring supports, body mounts, or various braces). While these components are mounted flush to each other, there is often some sort of space where moisture can accumulate and then begin to rust. Most of the undercar structural components are of heavier gauge steel than the sheetmetal body, so it takes them longer to rust. Also, they are typically out of sight, therefore out of mind.

Dirt

Dust and dirt accumulating on your vehicle is not a good thing either, for a couple of reasons. What is in that dirt? If you live in a farm community or in an industrial or construction area, it will be common for dirt to accumulate on your vehicle. In a farm community, this dirt will most likely contain fertilizers and pesticides, none of which are conducive to maintaining a pristine paint job. In industrial or construction areas, various types of metal shavings may get ground into the dirt that accumulates as dust on your vehicle. As these metal shavings are exposed to oxygen and moisture, they begin rusting while on the outer surface of your vehicle. This will begin to eat through to the surface of your vehicle if left unattended.

Clogged Drain Holes

Automotive designers realize that automotive vehicles are really beasts of burden, being exposed to rain, snow, the elements, and, if lucky, several car washes in that vehicle's lifetime. Since much of the vehicle's body sheet metal is a series of overlapping and intertwined pieces, small drain holes are strategically placed in the sheet metal to allow moisture to drain, rather than become trapped. However, dust, dirt, small gravel, tree leaves, and who knows what else can find their way in and under your vehicle, just to end up clogging these drain holes. Once the holes are clogged, they simply become a trap for most anything, causing rust to form in the process.

Road Salt/Sea Salt

As if moisture and oxygen are not bad enough to cause metal to rust, the presence of salt when added to the mix speeds up the formation of rust due to electrochemical reactions. This salt is present in areas that are located along saltwater coasts or where salt is used to treat roads and highways affected by snow and ice.

Humidity and Rain

Moisture from humidity and rainwater are constantly attacking the metal in your vehicle, whether it is the body's relatively thin sheet metal or the thicker chassis components. Any metal that is exposed to moisture, but not protected by paint, will begin to rust.

Whenever moisture gets trapped on a piece of bare metal and cannot run off, it will typically turn to rust before it evaporates. For this very reason, washing and then applying a good coat of wax to your vehicle is probably the best protection against the formation of rust on your vehicle's body panels. A good application of wax would help moisture (rain, dew, or even car wash water) to run off the surface, rather than sit anywhere the paint might be thin or damaged.

Body/Paint Damage

While the chassis components are somewhat out of sight, and thus out of mind, they are subject to road debris, gravel, and whatever else you may run over while driving. Even if a bouncing rock doesn't do any visible damage to your vehicle, if it breaks the surface paint of a chassis component, that is now an easy entry point for moisture and oxygen to begin forming rust. And if that paint chip is under the car, you may never notice it, until it becomes an issue.

Just like the underside of your vehicle, rock chips, scratches, and even small dents can break the paint surface in other areas of your vehicle. Any break in the paint surface can become the beginning of rust damage if left unrepaired.

The brown and white spot on the corner of the right front fender is where something has rubbed against the fender and scraped the paint off. If left alone, this small spot could be the beginning of serious rust damage.

COMMON LOCATIONS

Most any automobile can become a victim to rust if left out in the weather unprotected for long enough time. Sitting outside, rather than inside a garage; being neglected and never receiving proper washing and waxing; or being abandoned after an accident will put vehicles on the fast track to forming rust somewhere.

However, most all vehicles will have a weak or vulnerable spot, simply due to the design of that specific vehicle. There are also many common places where rust will take root and blossom on virtually every automobile if given enough time. Sadly, for some, it takes little time. A common joke about the Chevrolet Vega during the early 1970s was that they were known to rust while on the showroom floor. So where are these common areas where rust accumulates?

Rocker Panels

Rocker panels, the area below the doors, are notorious for rusting out because the rocker panels are often mistreated, both inside and outside. Any dirt or moisture on your shoes or boots when you get into your vehicle is often knocked off onto the rocker panel. Even though a metal or plastic trim strip typically covers the edge of carpeting or rubber that covers the floors, moisture or dirt can still get past that and begin eating away at the inside of the rocker panels.

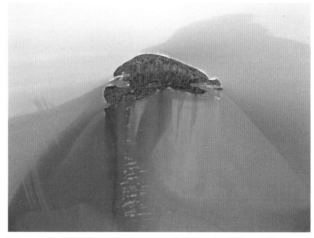

This vehicle obviously received some sort of impact, which could have been with another vehicle. More importantly, whatever caused the impact broke the paint surface. That allowed rainwater and other moisture to pool on the side of the vehicle and cause rust. The metal still looks fairly solid (not flaking apart) so the panel could probably be straightened and repainted if the vehicle's owner was so inclined.

On the outside, road debris, water, slush, and snow are kicked up by the front tires, providing a continual barrage onto the outer rocker panels. Because rockers panels are located low on the vehicle and often curve inward, they are not as visible and rust is often not seen until it has reached an advanced stage.

Other than the fact that this rocker panel has some dirt on it, it doesn't look too awful. However, it is possible that moisture could have worked its way under the edge of the carpeting, enabling the floor panel to begin rusting away. You may never know, unless you look beneath the carpet.

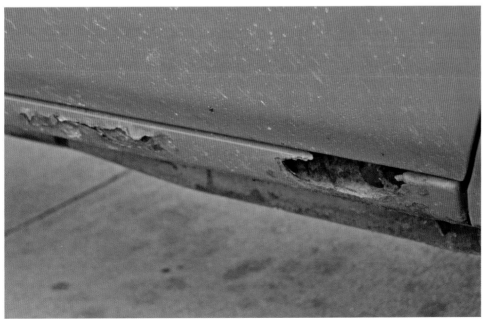

Although they are somewhat difficult to see, the specs of discoloration on this truck door are actually rock chips in the paint. While the door still looks solid, the rocker panel has fared much worse. Most likely, the driver of this vehicle lives or works on a gravel road. Mud flaps, a good coat of wax, and regular washing could have prevented at least some of this.

Rust is so common on rocker panels that new rocker panels are widely available for most collector vehicles through aftermarket sources. While the installation procedure will vary from one vehicle to another, a complete rocker panel replacement might be the best and easiest repair you can make.

Floor Panels

Much like rocker panels, floor panels receive abuse from both sides, inside and outside. While you may not be able to control what assaults your vehicle from underneath, drivers and passengers are often at least partially guilty for the floor panels' demise. Everything from salty French fries, soft drinks, baby formula, cigarette ashes, to anything else that you might spill inside your vehicle is probably not good for the floor panels.

An additional detriment to the floor is that it is often covered by carpeting or a layer of cheap insulation material. If liquid is spilled onto this material, it never really dries out. The moisture just stays concentrated on the floor panel at that one spot.

Around Headlights

Moisture and debris getting thrown from the front tires into an area that most likely never gets cleaned is the reason that many vehicles develop rust around the

Most likely, a rubber floor mat covered the floorboard of this truck for most of its past life. Moisture and dirt become trapped below this mat, causing surface rust at the least and complete rust-out at the worst. Since it "was just a truck," it was obviously not as cared for as the family sedan, judging by the gash in the metal near the middle of the photo.

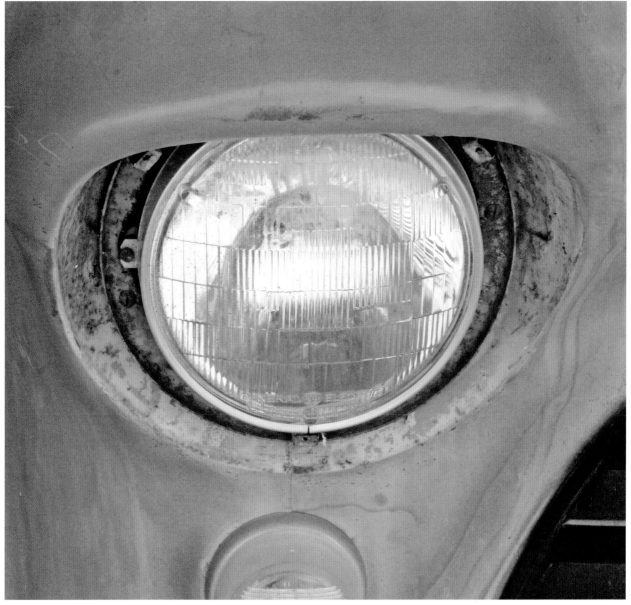

With the headlight trim removed, it is easy to see that the sheet metal that secures the headlight bulb is grungy. Being out of sight, this area is usually out of mind, so it never gets cleaned. Worse yet, the inside of the fender usually catches more dust, dirt, and moisture that is kicked up from the tires. The sheet metal directly above the headlight seems to be the most susceptible to rust.

headlights. The fact that this area is typically partially blocked off by sheet metal or rubber flaps makes it all but impossible to clean, if you were so inclined. There usually isn't any sort ventilation from the front of the vehicle into this abyss, so moisture that gets there, just stays there until it turns to rust.

Wheelwells

The wheels on the car go 'round and 'round, spraying and splashing everything on the roadway onto the wheelwells. Everything includes water, mud, dirt, gravel, road salt, sand, and trash of all kinds. For this reason, the wheelwells are doubly endangered. Hard objects such as sand, gravel, loose asphalt, broken glass, and nails can chip, scratch, or otherwise penetrate the paint or undercoating that exists in the wheelwell area. That is the first assault in the formation of rust, wherever it occurs. Secondly, rainwater, snow, dirt, and road salt get thrown in on top of the wound, forming rust in an area that most of us don't pay any attention to.

ABOVE: This is my wife's grandfather's truck that he pretty much ran the wheels off of before he passed away. Taking it to the car wash was unheard of, as it was rusty in ALL of the usual places. Wheelwells are one of those typical places for rust to accumulate, especially when some sort of trim is there to help trap dirt and water.

RIGHT: This shot of an inner fender wheelwell gives some idea how dirt and debris gets spread by the tire on a regular basis. Any of this dirt and moisture that gets trapped where these various pieces of sheet metal come together is sure to cause rust to form eventually.

ABOVE: Big off-road-type tires are going to throw more mud and, if the mud is not cleaned out, cause more damage.

LEFT: Judging by the rust on this vehicle (it was rusty in normal and non-normal locations), I would suspect that it has quite a story behind it. Although the paint still looks fairly decent between the door and the wheelwell, I would suspect that it is very soft.

Quarter Panels

While all of that road debris is attacking the wheelwells (the area between the inside of the tire and the vehicle's body), the same debris is also attacking the insides of the quarter panels (the sides of the car behind the doors). Front fenders are also just as susceptible for the same reasons. Don't think for a minute that pickup trucks are safe either … even though they don't have quarter panels—they do have bedsides. Whether you are discussing front fenders, quarter panels, or bedsides, having any sort of trim—whether it's chrome, stainless steel, or even a composite fascia—will aid in the accumulation of rust-forming material.

Trunk Floors

Just like floorpans, trunk floors are very susceptible to rust; however, much of it is caused by the vehicle owner/user and their amount of disrespect for the trunk. How many times have you seen the inside of someone else's car's trunk and it is full of crap? True, the trunk is designed for carrying our stuff while we are mobile. But that doesn't mean the trunk should be abused as much as it is. Concrete blocks, rakes, shovels, chains, floor jacks and other automotive tools, and broken bags of premix concrete are just some of the things that have done damage to the inside of a trunk floor. Lots of times, these items are simply thrown in the trunk without being secure, so they are just bouncing around.

How often does the trunk serve as the food table or snack bar for a picnic or other social event? No big deal right? And how many not-quite-finished beverage cans, cups, or bottles empty themselves on the trunk floor, but don't get mopped up?

Combining this abuse with the natural assault from Mother Nature and the trunk floor can be every bit a victim of rust as the floorpans. With the added concealment offered by the location of the gas tank below the trunk, rust can go unnoticed for quite a while.

Window Frames

As you might imagine, window frames and the body area around the windows are notorious for rust. Not so much on side or door glass, but on windshields and back glass, where the sheet metal that meets the glass has more contours. Because of the specialized shape, a somewhat thinner sheet metal is used to make relatively tight bends. As the glass is set into place, there is either

Although double-wall construction helps to prevent moving cargo in the pickup bed from causing damage to the outer bedsides, two adjacent panels make a great place for dirt and moisture to gather and create rust. Dirt and moisture gets thrown up by the rear tires, landing where it cannot escape.

Anyone who has owned a pickup truck for any length of time can probably attest to the fact that cab corners rust out quickly. The small space of a pickup cab can heat up quickly with a good heater on the coldest of days and can cool off quickly with a good A/C unit, a process that is bound to cause some humidity. That moisture has to go somewhere, so it will typically settle in the lowest spot on the cab. Add some dirt and grime, and hello rust.

What you see here is not only the rusted-through outer cab corner of a pickup truck, but also that the inner portion of the lower cab is pretty rusty now too. Once rust starts, it never sleeps, eating up whatever metal is in its path.

a rubber molding that wraps around the glass to make it weathertight or a type of adhesive that the glass is actually set into. This is what keeps wind and rain out, but it is often covered by some type of trim. So if this molding is ever damaged, the seal is thereby broken. Often, glass will need to be replaced because of an accident, rock chip, or hail damage. If the original seal and or adhesive is not removed completely, it will cause a void in the new seal. This then allows rainwater into an area that is typically hidden by the aforementioned trim. And that hidden moisture will turn to rust. This type of damage can be repaired, but to do so includes removing the glass, cutting out the damaged portion of the sheet metal, and then forming and welding in a patch panel. That is not impossible to do, but it is beyond the scope of this book.

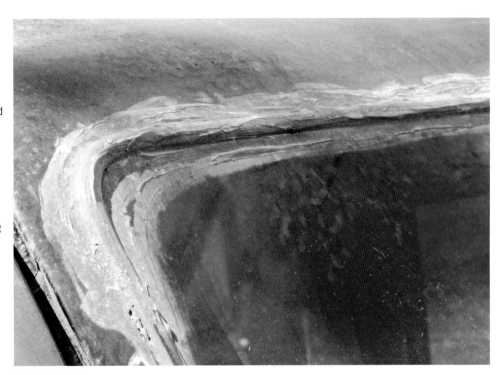

No rust in this photo, but what about beneath all the caulking? This caulking is a poor attempt at sealing the windshield frame. Most likely, the seal around the windshield was broken when the windshield was replaced at one time or another, but not resealed properly. An inadequate seal allowed moisture to gather around the sheet metal that makes up the windshield frame, causing that frame to rust. Once that happens, a windshield frame rebuild is about the only thing that will prevent leaks.

TYPES, SEVERITY, AND SYMPTOMS

Rust never sleeps. Once it begins, it is never going to stop, and it isn't going to go away on its own. You don't have to fix it, and on some vehicles, the amount of work to remove and repair hidden rust is simply more than the vehicle is worth. However, some rust damage is right there in your face. You see it every time you drive your vehicle, and you want to fix it. If you are going to repair rust damage, early detection is the best thing you can do.

Paint Scratches and Rock Chips

Some rust is relatively easy to repair. You just have to decide if you want to fix it or live with it. Nobody wants to be driving around in a vehicle that is primer spotted like a leopard. But, which is worse, having primer spots (or paint that just doesn't quite match) or having rust damage in various spots?

If your vehicle receives paint scratches, rock chips, and door dings, these are easy enough to repair. Most require only some touchup pain, some very fine sandpaper, and a little bit of time. In this situation, the damage is not deep enough to distort the metal, just deep enough to break the paint. A step-by-step procedure for making this repair will be shown later on in the book. Basically, you are just filling the scratch with paint, smoothing it out along the way, and matching the color as close as possible.

Smooth Surface Rust

Although there are certainly several degrees of rust damage, there are really only three when it comes to repair: surface rust, bubbled rust, and complete rust through. The first step has to be completely and accurately assessing the damage before you quickly get in over your head.

The easiest rust to deal with is surface rust. The sheetmetal body surfaces may simply be discolored (gray, brown, or reddish brown), yet still fairly smooth, with all body lines still distinguishable. This type of rust damage is common to a vehicle that has been left outside in the relatively dry elements for a lengthy time. Thin or well-worn paint no longer provides protection to the sheet metal, but wind and sunshine have kept it dry. Most likely, the rust damage has not been sped up by the presence of salt.

Surface rust is typically caused by something removing the paint from the surface and then the vehicle being left in a relatively dry climate. The dryer the better. This is something that hot rodders and auto restorers deal with often. Someone may have started stripping paint prior to repainting a vehicle and then lost interest or for whatever reason didn't finish the job. Perhaps the vehicle sat in a field of sandy soil for a time and the wind and sand combined to sandblast the paint off the vehicle.

True surface rust is nowhere as bad as it may look. The bad part is determining just where the line is between surface rust and damaging rust. While surface rust does alter the surface somewhat (microscopically by comparison), if it doesn't alter the shape and size of the affected panel, it still isn't a big deal.

A little different, but still relatively easy to deal with, is pitted surface rust. This type is common where

There was a time when pickup trucks were used exclusively for hauling loads, which meant that the load often was drug over the sides and tailgate without mercy. Repeated activity such as this will eventually wear the paint off, leaving the sheet metal below unprotected from the spread of rust.

A closer look at a rocker panel shown earlier. With the complete rust through at one location, it causes for speculation of some previous body damage at that area that may not have been repaired properly. Judging by the small rock chips in the door, it may have possibly been some larger piece of roadway debris that caused the initial localized damage, as the rest of the rocker panel looks fairly solid.

there is a bit more moisture and the surface has become pitted, much like a golf ball. The pits are not going to be very deep, simply because the sheet metal is not really thick. Regardless of the thickness of the sheet metal, the pits do not go through completely.

Repairing surface rust can be done by a number of methods. Choosing the best solution for you depends mostly on how you want to use your vehicle, how much work you are willing to do, and how much you can afford to spend … and the size of the area that is damaged. How you choose to address your impending repair depends largely on your automotive repair experience; the required tools, materials, and working space; your available timeframe; and the available budget. Most of these repair methods are somewhat easier if the affected panel can be removed from the vehicle. In these cases, a professional quality repair can be done without much more effort than some slipshod repair.

If the damage is limited to surface rust and/or road rash (rock/bug chips, and minor scratches), the repair strategy would be to remove any loose rust, get the surface clean, prime, and paint. If the surface rust is pitted, the repair would also require a skim coat of plastic body filler to fill the pits prior to painting. Anytime that body filler is required, that includes block sanding as well.

Removing surface rust can be done by a variety of chemical methods, by media blasting, or by sanding or grinding by hand. Each of these processes are described, in more detail, in Chapter 3. The better way would be by commercial chemical stripping or media blasting, but those methods are more appropriate for full restoration or custom builds. They do a complete job, work quite well, and are discussed elsewhere in this book. However, for the scope of this book, using paint stripper available at your local auto parts store or sanding with an electric sander or by hand would be more feasible.

While patina has been a catch phrase among hot rodders for some time to avoid using the term "rusty," it doesn't change the condition. However, the rust on this '67 Chevy pickup is certainly not a worst-case scenario. The sheet metal is solid and fairly straight, making it a great candidate for a restoration or custom project.

This 1950s Nash Metropolitan appears to have a fair amount of surface rust on it, yet still looks solid. Some of the rust may be more severe than the rest, but most of it would probably clean up pretty well. The hood is small enough to be stripped using paint stripper, followed up by some sandpaper, and then primed and painted.

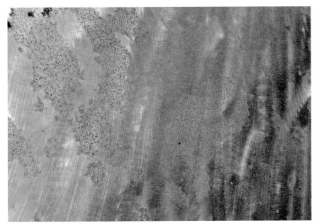

This is a closeup of a hot rod that was void of any paint at the time of the photo. The bare metal looking very smooth in some areas leads me to believe that paint and rust was removed by having the sheet metal commercially dipped. The brown areas are areas of very fine surface rust. A couple coats of filler primer would fill in these pits.

Regardless of which method you use to remove rust, if you go down to bare metal, you really should use epoxy primer to seal the metal before applying paint. If you do not have access to a spray gun (this book presumes that you do not), it might be worth your while to have a body shop spray your bare metal parts in epoxy primer for you. The epoxy primer will be considerably more durable in the fight against rust returning than any spray can primer would be.

After removing rust and any damaged paint, apply a skim coat of plastic body filler to fill in any pits left by the rust. After sanding the filler smooth, primer and paint can be applied.

Minor Surface Bubbles

When the surface begins to bubble, that is an indication that a localized area has been infected by rust. This is

During the 1980s, waterborne automotive paint was in its infancy, but it was still being used by domestic automotive manufacturers. Inconsistent adhesion between primers and top coats led many vehicles to have paint that would actually peel off in sheets. Surprisingly, the sheet metal below the beltline of this vehicle still looks to be intact. Some good sanding, proper primer, and paint would work wonders on this vehicle.

caused by moisture getting into the backside of the vehicle's sheet metal and causing the paint to separate itself from that sheet metal. At this point and anywhere beyond, repairing the rusty metal requires removal.

Removing rusty sheet metal can be done with tin snips, a cutoff wheel on a die grinder, or a reciprocating metal saw. When cutting sheet metal, you must always be aware of sharp edges and the possibility of sparks, therefore mechanic's gloves and eye goggles are highly recommended.

Prior to cutting away original sheet metal, it is always a good idea to use a marker or tape to outline the area that you plan to cut. While you are concentrating on cutting, it is sometimes easy to lose track of where you are in the process and actually cut away more than you intend to, especially when using a powered device. There is no need to create more work for yourself.

Since bubbled rust means that the sheet metal is actually deteriorating, this metal will require replacement.

The affected portion of the panel should be removed and replaced. The correct way to do this would be to weld in new sheet metal, whether it is a commercially available patch panel or one of the hand-crafted variety. However, since welding is beyond the scope of this book, alternative methods are rebuilding the area with fiberglass or using panel bond adhesive to secure a metal patch panel in place. While it is relatively new technology to many people, panel bond adhesive has been used in automotive manufacturing and collision repair for several years.

Severe Bubbles

If the previous condition of minor surface bubbles is left untreated, the bubbles get more pronounced and the affected area increases in size. The method of repair is essentially the same, although you will typically be repairing a larger area. The real downfall of this is being able to duplicate the original body lines and contour that are involved over a larger area.

It looks like road rash chipped the paint on the hood of this older vehicle and the paint was not touched up, allowing rust to form. The rust then began to bubble up from below the paint, causing bubbles in the surrounding painted surface.

Multiple types of rust in one shot: small bubbles of rust, large bubbles of rust, delamination of sheet metal and paint, along with complete rust through. Again, with rust damage this severe, I would suspect some type of collision damage in the area of this extended pickup cab.

COMPLETE RUST THROUGH

Again, any time that the rust is allowed to become more than simple surface rust, the repair involves removing the rusty metal and re-creating the original sheet metal with sheet metal or fiberglass. Whether it is bubbled rust or complete rust through, a concern with repair is how far does the rust really go, how much original sheet metal must be removed, and how difficult is it going to be to reproduce the various shapes that might fall within that area.

As long as you have something to serve as a framework, fiberglass cloth and mat can be formed into most any shape. After fiberglass is used to repair the structural integrity of the missing sheet metal, plastic body filler can then be used to make it look pretty. However, sheet metal (thick enough to use in auto body repair) is going to require more skill to shape into anything other than basic shapes without the use of sheetmetal equipment.

ALTERNATIVE TO REPAIR

If all of this has your head swimming in doubt of your abilities, there are other options, depending on the location of the rust. Most any sheet metal on the front clip of your vehicle—the front fenders, hood, or inner fenders—can be replaced. These panels all bolt together, so with a set of wrenches, they can be removed and replaced. Perhaps the most difficult part of doing this will be finding that "one last bolt" that is preventing you from removing the damaged panel. Having access to an assembly manual for your vehicle will help with this.

If rust is in the doors, they can be replaced also. Depending on the vehicle and the type of hinges and latches, this might be a little more difficult. Additionally, there will be door glass and potentially interior upholstery to contend with.

Except for the deck lid, most anything behind the doors is not going to be replaceable—other than simply obtaining a different vehicle.

Replacing Panels

Going the replacement route can be an interesting trip, depending largely on what vehicle you are working on, its vintage, and its rarity. One of the great things about most automobiles built within the last fifty years is that hundreds of thousands of them were built. This means that probably the fender or door that you need are out there … somewhere.

Rust through on an edge is probably more difficult to repair than in the middle of a panel. If the rust through is surrounded by good metal, new metal can be riveted or welded in place and then the area filled with plastic body filler as necessary to rebuild the necessary shape. When the damage is on the edge, recontouring the edge can prove to be tricky.

Again, multiple layers of sheet metal lapped together makes for a perfect breeding ground for rust. Some of the rust still looks solid, yet it is adjacent to areas of complete rust through, so it may be soft if you were to start poking around on it. Whenever rust occurs on multilayered panels like this, the repair might be too complicated for the weekend rust warrior.

Salvage yards used to be the perfect place to find most any automotive parts that you wanted, but those yards are getting fewer and farther between these days. Automotive swap meets, the bigger the better, usually have a wide variety of parts, so those are a good source.

OEM and Aftermarket

Although buying new parts is typically going to be more expensive than from a salvage yard, doing so would typically require less work. Still, that does not translate into no work. Regardless of the material of new parts (steel, fiberglass, or carbon fiber), they will still require priming and painting regardless of who made them or where they were purchased. What might not be quite so obvious is that even though they may be advertised as direct replacement parts, they may not fit properly. If this is the case, the time and effort spent getting the new part to fit properly might be more than what would have been required to repair the original.

Poor fit is not always the case with new parts. However, it is something to be aware of during your search for a replacement part. The good news is that replacement parts typically come primed with baked-on primer from the factory. Once you get them to fit, they can be scuffed with an abrasive pad and painted in a short amount of time.

Used

Just because a vehicle part is used, that doesn't mean that it is used up. Salvage yards and private parties often have sheetmetal parts that were not damaged in an accident that otherwise totaled a vehicle. Or perhaps a vehicle's mechanical issues were simply beyond economically feasible to repair. While a part from another vehicle similar to yours may not fit exactly, coming from the same manufacturer should make it fit somewhat better.

Whether you purchase new or used, realize that the replacement part or parts may not include any trim, emblems, or mounting hardware. To avoid trying to find some hardware that is no longer available, hang onto your original parts until after your repair or replacement is completed.

Chapter 2
Repair Equipment

TOOLS (COMMON)

For most any task, certain tools are required to do the work correctly. Sometimes, a substitute can be used, sometimes not. For repairing relatively small areas of rust on an automobile, you may already have some of these tools in your garage if you are the slightest bit of a do-it-yourself kind of guy or gal. We are not talking about doing collision repair or even professional quality auto restoration, so the list of necessary tools is short and they are not very expensive. Most of what you need can be purchased at your local automotive parts store (NAPA, O'Reilly's, AutoZone, etc.). You can also find most of what you need at discount department stores or home improvement stores.

At a minimum, which may be adequate for your needs, you will need wire brushes and tin snips for removing rust. To make small patch panels, you will need a body hammer or two and a body dolly. You will also need a pop-rivet gun to attach the new sheet metal to the old, along with a sanding block to sand everything smooth.

If you already have an air compressor, you can save some elbow grease by using a cutoff wheel in place of tin snips. You can also use spray guns for applying primer and paint. However, for now we are going to assume that you do not have an air compressor and are going to use spray cans for primer and paint. If the project at hand requires better results than what you can obtain with spray cans, I must refer you to a couple of my other books: *The Complete Guide to Auto Body Repair* and *How to Paint Your Car*, both of which are available from Quarto Publishing, as part of the Motorbooks Workshop series.

Wire Brushes

Wire brushes are available as a piece of contoured wood with wire bristles protruding from it or as an attachment for a drill motor or grinder (pneumatic or electric powered). If you are unsure as to the severity of the rust damage, you can usually find out pretty quick by using a wire brush to brush over the affected area. This will work to dislodge dust, dirt, and some surface rust quickly. If after a few swipes with the wire brush you realize that the metal below is now becoming shiny, the situation might not be as bad as you thought. If the metal below is shiny and shows no pits or holes, it may not require any rust repair at all, just primer and paint.

If the shiny metal does show pits, you may need to fill the pits and then prime and paint, but only after taking additional steps to evaluate the condition of the metal. If the pits have turned into holes, yes there is rust, and you now need to determine its severity. For these exploratory steps, a handheld brush is quite adequate and is very affordable compared to purchasing a drill or grinder. You can probably purchase a handheld wire brush for less than $5 or perhaps a set of two or three different sizes for less than $10.

Tin Snips (Aviation Snips)

Compound leverage snips, aviation snips, or tin snips are all common names for what I will refer to as tin snips. These are available in different styles, prices, and material capacity. For automotive work, you will typically be working in the 18–22-gauge metal thicknesses.

Small handheld wire brushes like these are good for doing preliminary investigation of formations of rust. Brushing away any remaining paint will give you a good idea as to the depth of the rust. If it isn't bad, applying some primer and new paint might be sufficient.

If after doing some investigation, you realize that your rust repair is going to involve some rust removal and bodywork (body filler, primer, and paint), some mechanized wire brushing will make getting down to bare metal easier and quicker. When using any type of wire brush in a power tool, be sure to wear gloves and eye protection.

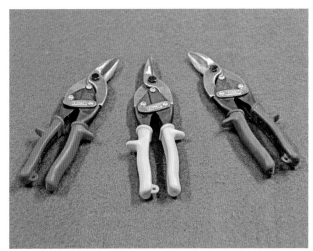

After cleaning the rust out of the way so that you can clearly see its extent, aviation snips can be used to cut away some of the cancerous sheet metal. These are also used to cut flat sheet metal or aluminum sheets into smaller pieces that can be riveted or welded into place to replace what you cut out. While the yellow-handled snips cut straight, the green handles cut straight or to the right and the red-handled snips cut straight or to the left.

These snips do not provide the leverage of compound leverage snips, so they can be a little harder to use, but that is really whatever you get used to. However, these are better at cutting straight edges (rather than curves).

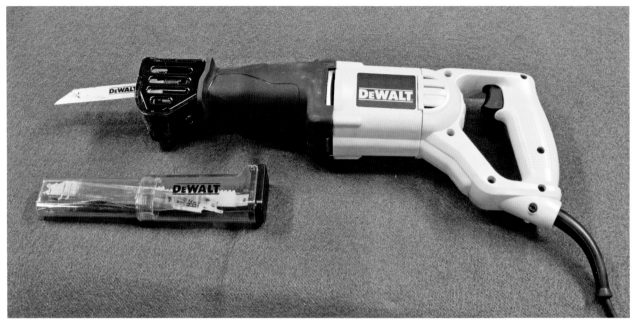

A reciprocating saw such as this one can sure make quick work of cutting out rusty sheet metal when equipped with a metal cutting blade. This one is powered by electricity, but pneumatic units are also available if you have access to an air compressor. Different blades are available for cutting wood or metal, making this a handy tool to have at your disposal.

RECIPROCATING METAL SAW

While most reciprocating metal saws designed for automotive use are pneumatically operated and therefore require an air compressor, many homeowners now have the same type of equipment that are electrically powered. When equipped with a metal cutting blade, these work every bit as good as one designed for automotive work. The only foreseeable drawback is that one from a home building supply store will be significantly larger, possibly limiting its maneuverability. However, if that is not a problem, bigger is usually better and sure beats using tin snips.

BODY HAMMER

Leading the list of basic tools are hammers, and a variety of hammers are used for bodywork. They all have their specific characteristics and applications. There are situations that call for a big hammer and others that call for a smaller hammer and a lighter touch. Learning and knowing the difference between the two is something that amateur auto body workers often have trouble with.

Most body hammers have a head and a pick, making each hammer a dual-purpose tool. The head is usually large (between 1 and 2 inches in diameter) and relatively flat with a smooth surface, while the pick

end is much smaller and pointed. The larger head is used for flattening metal against a dolly. The pick end is typically used for hammering out very small, localized dents, with or without a dolly. Picks can come to a very narrow point or to more of a blunt point.

When sheet metal is bent in a collision, in addition to bending, it stretches. To reduce some of this stretching, a shrinking hammer is used. Shrinking hammers are similar to other hammers, except that the head is serrated.

Different hammer manufacturers combine different heads with different picks. When buying bodywork hammers, having a set that contains a flat face, a shrinking face, a blunt pick, and a sharp-pointed pick will fulfill most of your hammering needs.

In regard to rust repair, hammers and body dollies will be necessary if you have damage that is severe enough that requires metal replacement. Unless you purchase a vehicle specific patch panel, or a comparable piece of sheet metal from a donor car, you will be required to shape a piece of flat sheet metal or aluminum stock into the basic shape of the original sheet metal that is now missing. For this specific reason, if you cut out any stock sheet metal, try to cut it out in its entirety, so that you have a better idea of what you need to duplicate.

For auto body repair, a large selection of hammer shapes are necessary because automotive sheetmetal panels have lots of shape to them. For rust repair on your daily driver, you certainly do not need all of these. If you had only one body hammer, the one shown in the lower left would be the most universal. It has a good size flat surface on one end and a rounded, pointy end on the other. The small end might be handy for determining how rusty a panel is and for creating a narrow bead on a replacement panel.

I recently purchased this seven-piece auto body hammer and dolly set from Eastwood. It includes three hammers, providing different smooth face heads and a shrinking head, along with a round pick and a chisel point. The three dollies included are a universal (which provides the most different surfaces), a heel dolly, and a toe dolly. At the upper right is a slapping spoon, commonly used for reaching behind a dented panel and slapping it outward toward its original contour.

Body Dolly

Made of hardened steel that has been smoothed, dollies come in a variety of shapes and sizes. Dollies are usually held on the backside of the metal being straightened, while a hammer on the outside flattens the metal between the two, resulting in metal that is roughly the shape of the portion of the dolly being used. For this reason, having a variety of dollies with small, large, convex, and concave shapes will add to the versatility. Like shrinking hammers, dollies with a serrated face will help to shrink stretched metal.

POP-RIVET GUN

To properly educate you and to keep myself out of trouble, I should probably tell you that *POP* in POP rivets is actually a registered trademark of Stanley Engineered Fastening. So when you go into a hardware store to purchase a sheetmetal fastening device, the store may have what you need, but it may be called something different.

Regardless of what brand you may use, if you are trying to fasten multiple layers of sheet metal or aluminum together without welding, rivets are a natural choice. They are available in different sizes, styles, and materials. At the risk of oversimplifying the riveting process, the panels (aluminum or sheet metal) are first formed into the necessary shape with some amount of overlap. This overlap is what allows for the use of rivets. With the panels clamped in place in their correct alignment, holes that correspond to the size rivet being used are drilled through each panel. The head end of the rivet is then inserted through the holes in the panels. This will leave a thinner and longer portion of the rivet protruding out toward you. Place the rivet tool (handheld or pneumatic) over this protrusion, and then as the rivet tool is operated, it squeezes the head end of the rivet flat against the backside of the panels, holding them together in the process. The rivet tool then also cuts off the thinner and longer portion of the rivet that is no longer needed. Please note that these scraps should be disposed of quickly, as they will go through an automotive tire very quickly if driven over.

Die Grinder, Angle-Head Grinders, and Cutoff Wheels

For making relatively straight cuts in sheet metal, a die grinder with a cutoff wheel works very well. Most commercially available patch panels have straight edges, so using a die grinder allows you to cut out a similar shape quite easily. Be sure to leave about a half inch of the old metal to overlap the patch panel and the new panel can be plug welded in place quite easily.

Rivets are made of different diameters, depending on the intended use. Regardless of size, the thinner, longer portion of the rivet is inserted into the rivet gun. This photo shows a hand-operated model that is adjustable to use two different sizes of rivets. The portion of the rivet that extends out of the rivet gun is what passes through the material to be joined. By squeezing the handles together, the rivet is squeezed together and the long shank is broken off.

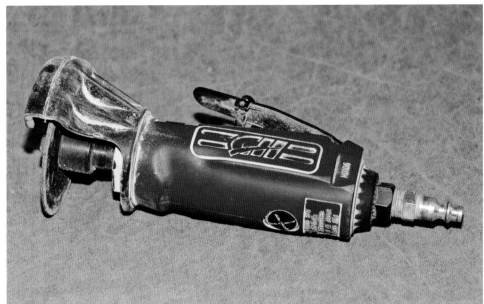

This pneumatic die grinder can be used to make straight cuts in metal. It is typically used as a cutoff tool for small material or to cut out old metal where a patch panel will be installed. Prices range from less than $50 up to about $100.

An angle-head grinder works the same as an ordinary die grinder, but with its angled head, it can sometimes be used in areas where an ordinary die grinder cannot get into, depending on how flexible the air hose may be. In addition to cutoff wheels and sanding discs, either of these tools can also be fitted with a wire wheel or a variety of other specialty grinding heads.

Although they operate the same and perform the same task, die grinders are available in two distinctly different configurations. In both types, the air hose attaches to the end of the die grinder's body, the body serves as the handle when in use, and a lever-type trigger is squeezed against the body to operate it. The difference in the two styles is that on one, the grinding wheel rotates perpendicular to the body of the die grinder, while on the other the grinding head is mounted at 90 degrees to the body. If you have plenty of room in which to work, this is not a big deal, but when space is limited, a die grinder with the angled head will usually be more maneuverable.

Sanding Block

Sanding blocks are commercially available in a very wide variety of shapes, styles, sizes, and materials, while improvised sanding blocks are pretty much unlimited. No matter how good you may be at straightening some damaged sheet metal, unless you exert the required effort to get the surface both smooth and flat, the finished paint simply will not look its best. Sanding is

Just as there are different types of sandpaper, there are different types of sanding boards and sanding blocks. The flexsand block in the middle and at the left have a hook-and-loop method to attach the sandpaper and are used for sanding slightly curved surfaces. The two wooden handled sanding boards are for sanding flat areas and clamp the sandpaper in place. The smaller rubber sanding block in the lower right is slotted on both ends and conceals tacks that secure the sandpaper.

Even with all of the commercially available sanding devices, some body contours call for improvisation. Short sections of radiator hose, fuel line hose, or a paint stick can be wrapped with the sandpaper of choice to get into those otherwise difficult areas. Just remember that a sanding block must provide a uniform backing surface to properly do the job.

what is required to achieve these elusive goals, while sanding without the aid of a sanding block is merely wasting your time.

MATERIALS

Most, if not all, of the material products associated with auto body repair contain some type of solvent or other chemicals that can cause skin irritation or worse if your skin comes in direct contact with it. For this reason, it is imperative to read the warning labels with these materials prior to using them. Even if the warnings do not suggest wearing disposable gloves when working with these materials, it is a good practice to get into. Disposable gloves are inexpensive and using them sure beats picking fiberglass resin or body filler from your fingers and cuticles.

Wax and Grease Remover

Before sanding, priming, or painting any surface, that surface needs to be as clean as possible. All traces of dirt, grease, oil, silicone, or other contaminants must be removed. If the surface is not clean prior to sanding, you run the risk of smearing any contaminants that are present onto a larger area and also further embedding those contaminants into the surface. If the surface is not clean prior to applying primer or paint, the freshly applied primer or paint simply will not adhere properly. Even oil from your fingers is enough to deter proper adhesion, so avoid touching the surface with your bare hands.

To obtain this clean surface, using wax and grease remover is a necessity. Wax and grease remover is readily available at your favorite paint supplier and is relatively inexpensive, so there is no reason to be without it. After using an air gun to blow away any dust or dirt that is on the surface, the wax and grease remover is applied with a clean cloth (or paper towel) or sprayed onto the surface. Wipe the surface dry, using a clean, dry cloth (or paper towel). It should be apparent that the use of a clean cloth is of utmost importance here as using any old shop towel that may have grease, brake fluid, or any other contaminants would be defeating the purpose.

Rust Dissolver Gel

Rust dissolving gel works on iron or steel to remove rust.

If possible, the surface should first be scuffed with a wire brush, and then the gel applied. After the gel sits for five to fifteen minutes, it should be cleaned away, and the surface neutralized with water. After removing the rust, the iron or steel should then be primed and painted to minimize the reoccurrence of rust.

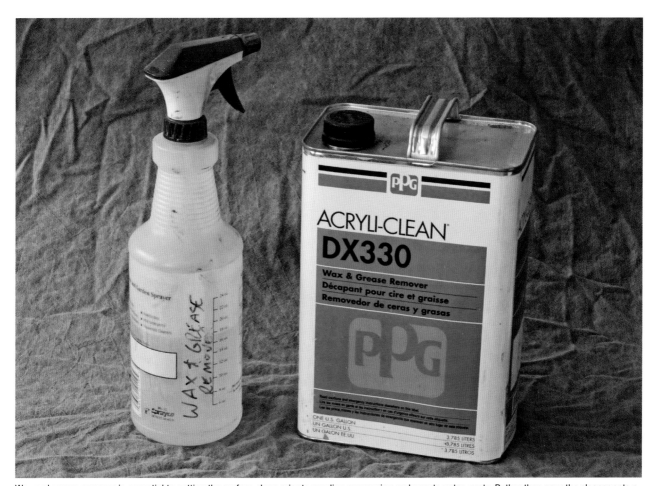

Wax and grease remover is essential to getting the surface clean prior to sanding or spraying undercoats or top coats. Rather than pour the cleaner onto a cloth or paper towel, invest in a cheap spray bottle to apply the wax and grease remover, and then wipe it off with a clean towel. Just be sure to label the bottle with its contents.

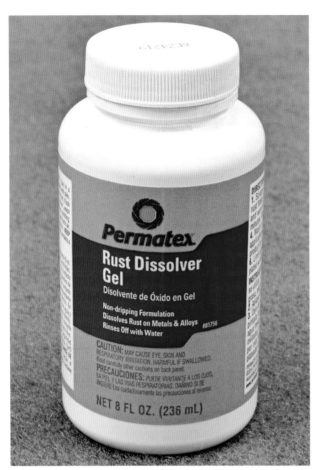

Rust dissolver gel can be brushed or sprayed onto iron or steel to remove rust. After application, the gel should be allowed to penetrate for five to fifteen minutes to remove loose rust. It must then be cleaned away, the area neutralized with water, and the area coated with the rust prevention material of your choice.

Paint Stripper

Old paint can be removed from steel parts and some other metals with paint stripper. This type of product may not be suitable for use on all metals or finishes, so you should consult your paint supply jobber for their recommendations on which products to use. A particular brand that is commonly used for stripping auto paint is Klean-Strip Aircraft Paint Remover. There are other brands available, however.

Contrary to what many people believe, chemical stripping does not involve acid. It does involve a collection of chemicals that you must be careful with during use and disposal though. Chemical paint strippers are safe to use, but you must respect their dangers and follow the appropriate safety precautions to prevent burns to your skin or other skin irritations.

Sandpaper

Prior to applying sealer and paint, you will get plenty of opportunities to work with sandpaper, so there is no need to start using it too soon. Lots of sandpaper will be used and lots of sanding done before the project is finished, but damaged body panels should be straightened long before any sanding is done. Sandpaper should be used for smoothing thin coats of body filler, scuffing a primed surface prior to application of additional coats of primer or sealer, and for wet sanding clear coats after the paint has been applied.

Although you may have heard of wet sanding, you may not know what it actually means or when it should be done. First, to use it wet, you must use sandpaper that is designed for and labeled as wet or dry. If you are using sandpaper that is not designed to be used wet, it will simply fall apart when it gets wet.

Wet sanding is typically done only after the vehicle has been painted. Extremely fine (1,000-grit or finer) sandpaper is moved in a circular motion with light pressure after being dipped in a water bucket or by spraying water onto the surface being sanded. The wet sanding process removes the orange peel effect from the paint, while the water helps to float away the paint that is being removed, rather than simply rubbing it back into the surface.

On some high-dollar, custom-built vehicles, the person doing the bodywork may use the wet sanding technique prior to paint in order to get the smoothest surface possible. For repairs to your daily driver, this is probably overkill and as such is a wasted effort. It is also not a good idea to be pouring water on a piece of bare sheet metal or into an area of body filler. Unless you are a highly skilled body person, you don't need to be creating any more problems for yourself while you are trying to do bodywork.

So now that you know that most of your sanding is going to be done dry, grab yourself a dust mask and a sanding block. For removing paint and getting down to bare metal, a 36- or 50-grit disc on an electric or pneumatic sander works best if you are working on a localized repair. Entire panels that require complete paint removal should be chemically stripped or media blasted. For initial shaping of body filler, use 80- or 100-grit sandpaper over the entire filled area and then switch to 200- or 240-grit sandpaper to blend the filler into the surrounding areas. The entire area that will require repainting after repair should then be sanded with 400-grit sandpaper.

When using any type of paint stripper, be sure to wear disposable gloves to protect your hands, along with a long-sleeved shirt and long pants to avoid chemical burns to exposed skin. You should also use cheap, disposable brushes and containers as cleanup afterward is nearly impossible.

Fiberglass (Resin, Mat, and Cloth)

For custom work or bodywork repair, fiberglass is a material that is easy to learn to work with and does not require specialized equipment. Fiberglass can obviously be used to repair components that are made of fiberglass, but it can also be used as an alternative to welding in sheetmetal patch panels. For repairing sheet metal, welding in metal patch panels provides the best overall repair. However, to do so will require a welder and patch panel, which are sometimes not available for the task at hand.

Fiberglass mat is commonly used when repairing holes or anytime where a bulk buildup is required. It is made up of short strands of glass fibers that are bound together, but not woven. When the mat is saturated with fiberglass resin, it becomes quite pliable and easy to form into complex shapes. As each layer of fiberglass mat is added, the laminate becomes thicker and stiffer.

Fiberglass cloth is more suitable to repairing cracks, breaks, or holes were there are no or few complex shapes. Since fiberglass cloth is made up of glass thread that is woven together, it is difficult to mold to curved shapes without wrinkling. To get the cloth to lie flat when it must be used on a curved surface, several small cuts can be made in the cloth before the resin is

Fiberglass mat consists of short strands of glass fibers, making it easy to form into complex shapes. Each successive layer of fiberglass mat and resin makes the laminate thicker and stiffer.

Fiberglass cloth is woven glass fiber threads, making it stronger than fiberglass mat of the same thickness. Cloth is used more for repairs, while mat is used to build bulk.

applied. Fiberglass cloth is thinner than fiberglass mat, so cloth should not be used when you are building up bulk in the repair area. However, a repair made with cloth will be stronger than an equal thickness repair made of mat since the cloth will contain a higher percentage of glass.

Fiberglass mat and cloth is available in different widths, lengths, and weights for different size jobs. For the beginner, working with smaller pieces of cloth or mat will be easier than trying to work with larger ones.

To bond fiberglass mat or cloth, use fiberglass resin. Resin is a two-part mixture comprised of a thick liquid, to which a much smaller amount per volume of hardener is added. The amount of hardener to use will vary depending on the ambient air temperature and humidity, and the amount of working time you desire when working with the material. As the fiberglass resin cures or "kicks," a chemical reaction is occurring between the liquid resin and the hardener. As a result of this chemical reaction, two things happen: the fiberglass resin, mat, and cloth solidify and they get very hot in the process. For this latter reason, adequate skin and eye protection is mandatory when working with fiberglass. Adequate ventilation is also a necessity as the odor of curing fiberglass is quite intense and can be quite irritating to those with respiratory problems. After the fiberglass has cured, you should also wear a nuisance mask, eye protection, and skin covering anytime that you are sanding, drilling, or cutting it.

To estimate how much resin you will need when making a repair, you should first remember that each of the second and successive coats requires only half as much resin as the first coat. For the first coat, it will take approximately one pint of resin to thoroughly saturate one square yard of cloth, while 1 ½ quarts of resin will be needed to thoroughly saturate one square yard of mat.

When mixing and working with fiberglass, use materials that are disposable rather than cleanable. Fiberglass resin can be mixed in disposable paint tray liners, picnic/party cups, empty gallon milk jugs, or most anything that is clean and will contain the liquid resin. Common paint stirrers work well to mix the resin and hardener, while disposable paintbrushes work well to spread the resin onto the cloth or mat. With all the disposable containers available, they are cheaper to replace than the materials that would be required to clean them, not to mention the time savings. Your hands, however, are not disposable, so cover them with disposable latex gloves. You should also have some acetone available for cleanup should you happen to get some resin on your hands or skin.

Sheet Metal

Most large cities have metal supply retailers that sell sheet metal in large sheets, as well as structural steel in a variety of shapes and sizes. If one of these outlets are convenient for you, they will most likely have a wider variety of products to choose from and will ultimately

Fiberglass resin, when mixed with hardener, is the catalyst that holds the fiberglass mat or fiberglass cloth together. When mixing, use cheap disposable containers, mixing sticks, and trowels that can be thrown away after use.

be less expensive. However, if you require only a small piece of sheet metal for your automotive repair, your local hardware store may be more convenient for a one-time repair.

As mentioned previously, most automotive sheet metal is between 18- and 22-gauge thickness, with the smaller number indicating a greater material thickness than the numerically higher number. Using sheet metal that is thicker than 18 gauge will be more difficult for a novice to shape correctly, without the aid of specialized bending and shearing equipment. Anything thinner than 22 gauge will usually not be strong enough for use as an automotive sheetmetal patch.

While a metal supply house will have sheets of metal as large as 48 inches by 96 inches, the local hardware store will typically have smaller pieces that are easier to transport.

Panel Bonding Adhesives

Perhaps one of the biggest developments in automotive bodywork recently is the development and use of panel bonding adhesives. As automotive manufacturers have worked to decrease the overall weight of their vehicles, they began using body components that were made of composites, rather than steel. Since these composite materials cannot be welded using the same methods as welding steel, bonding adhesives were developed. Taking a good thing even further, panel bonding adhesives have now been developed for use with steel panels.

Multiple companies produce panel bonding adhesives and different products are available for different situations, so be sure to read the package labels closely to ensure that you purchase the correct product for your application. You can buy panel bonding products for assembling rigid composites and for flexible products. They are used in the same basic manner, but using the correct material is the key to achieving the best results.

Most, if not all, panel bonding adhesives are a two-part epoxy. With improper mixing being a common source of failure for any type of two-part epoxy, manufacturers have developed a system that mixes the two parts as they are used, rather than relying on the user to measure and mix thoroughly. The epoxy components are packaged in a tube, much like caulking that would be used for sealing around household windows or bathtubs. Each of the two components is inserted into a specialized application tool, somewhat like a caulking gun. After removing air from both tubes, a mixing tube is attached to the tip of the application tool. As the trigger is squeezed, both components are blended together as they travel through the mixing tube. Typically, the adhesive is applied to both surfaces to be attached. The workpieces are then positioned and clamped into place. Most panels are cured in less than twenty-four hours, but with heat lamps or other controlled application of heat, the procedure can be sped up considerably.

While panel bonding adhesives will not completely eliminate the need for welding steel when doing bodywork, they have minimized welding substantially in both the auto manufacturing and collision repair industries. Not only is panel bonding a time saver, but it has been found to be just as strong and durable as welded panels when involved in a collision. It also eliminates the risk of panels being warped when subjected to significant heat when being welded by conventional methods. Panel bonding adhesives also are not a fire hazard. The downside of this type of repair is the cost of the application tool. In a collision repair shop, it will make up for itself quickly in reduced labor costs, but it might be out of reach for a hobbyist making a one-time repair.

Plastic Body Filler

No matter how good you may be with a body hammer, if you have straightened any sheet metal or installed any patch panels, you will probably need to apply at least a skim coat of body filler. True, some craftsmen can metal finish a vehicle so perfectly that no filler is required. However, you probably aren't to that point quite yet. Before applying any filler, you should take the time to consult with an expert to verify that you are using compatible materials. You will be much better off to determine the appropriate filler product for your use before you use it, rather than be disappointed when a less-than-optimum product yields less-than-pleasing results. The main concern here is whether you are applying the body filler to sheet metal, galvanized steel, fiberglass, or aluminum. It is always better to do the work correctly the first time than to redo a bunch of work or to be forced to live with less-than-desired results.

If you purchase your bodywork supplies at a dealer that supplies these same materials to the professionals, you can rely on the fact that the person behind the counter knows which products are best for your

Not all body fillers are created equal, as they are designed for specific applications. Just as when choosing paint materials, you should stay within one brand of fillers to eliminate compatibility problems. When you purchase body filler, make sure that you pick up the tube of hardener that goes with it.

particular application. However, if you purchase your body filler from the local discount department store, you may be hard pressed to find someone who actually knows what the product is used for. You can be certain that most body fillers will have labels indicating they can be used on bare metals; still, that does little to inform you of the appropriateness of the product on your particular project.

Plastic body fillers have evolved greatly since their inception as an alternative to lead. Several different companies now make plastic body filler, and most offer a variety of products to choose from, depending on your application. Some fillers are designed for use over fiberglass, while others are designed for use on sheet metal. Still others require an undercoat of epoxy primer to increase adhesion, while others should be applied directly to bare metal. Like various primer products, some fillers are for smoothing out rough bodywork, while others are used for finish coats. Quite simply, not all body fillers are created equal or even to perform the same task. If you use the wrong type of filler, it will quickly reveal the location of any bodywork when the vehicle is parked in the sun. I doubt that you plan to keep your car parked in the garage forever.

Seam Sealer

Seam sealer is much like caulking for automobiles. Although it is available in forms that can be brushed on, it is typically dispensed from a tube. As its name would imply, it is used to seal seams in sheet metal from moisture or dirt that would eventually allow rust to begin forming. Common areas of use include on floor panels and in trunk areas where the floor meets the inner fender or wheelhouse. Any location that is prone to collecting and trapping moisture is a good candidate for an application of seam sealer, as its application is much easier and less expensive than replacing rusty sheet metal. Most seam sealer products can be applied directly to bare sheet metal or over primered surfaces, but they are usually applied prior to paint.

Touchup Paint

Although it seems to be getting more difficult to find these days, automotive touchup paint is designed to be a perfect match for your vehicle, in a convenient package. Its primary purpose is to touch up or cover small breaks in your vehicle's paint, such as those from a small scratch or rock chip.

If your vehicle still has its original paint and the paint tag, you can find the paint code and then match it up at the parts store or wherever you buy touchup paint. Most comes in a small tube less than an inch in diameter and about 3 inches tall. The cap has a brush attached, much like nail polish.

After finding the appropriate color, you should shake up the bottle's contents, then use the supplied brush to gently and lightly apply touchup paint to the blemished sheet metal. Multiple light coats will be better than one heavy coat. While this will not fix a deep scratch, it will typically cover a light scratch that isn't very deep, and thus help prevent rust from forming in that area.

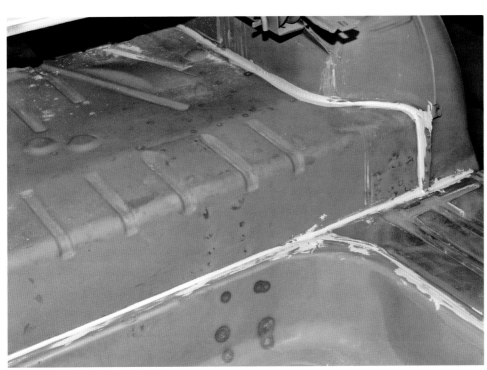

Original seam sealer that has gone bad is part of the reason that trunk floors fall to rust. Since several panels actually make up the trunk floor and surrounding area, there are several seams for moisture to collect in. A preventative action would be to remove the dried-out seam sealer when it becomes too dried out to work, but no one does that.

With the paint code from your vehicle, presuming that it is still sporting the OEM color, you can purchase touchup paint that should match your vehicle perfectly. With a brush attached to the inside of the bottle cap, it is perfect for touching up small rock chips or scratches. Doing this will cover any broken paint, preventing rust from starting.

products. Ordinary household masking tape has not been treated to withstand the potent solvents that are used in automotive paint. Additionally, adhesives used in ordinary masking tape are not designed to easily break loose from surfaces and can remain on painted bodies after the bulk of material has been pulled off. Lingering traces of tape and adhesive residue might require use of a mild solvent for complete removal, a chore that could threaten the finish or new paint applied next to it.

Whether your job consists of a very small paint touchup or complete paint job, you have to realize that automotive paint masking tape is the only product designed for such use. Using any other type of inexpensive alternative is just asking for problems and aggravation.

Automotive-grade masking tape is available in sizes ranging from ⅛ inch up to 2 inches wide. You will likely use ¾-inch tape for most purposes, but having a couple of extra sizes will make your masking work easier. It is much simpler to place a few strips of 2-inch-wide

Aerosol (Spray Can) Paint

Aerosol paint is similar to touchup paint in that you can sometimes find paint that is the same color as the OEM paint on your vehicle. Other times, you may be forced to settle for a color that is not an exact match, but is close enough.

Of course, spray paint is available in many different colors and is convenient to use, making it great for refinishing accessory items of all sorts. Several formulations of specialty paint, such as that used on OEM brackets, brake calipers, engine blocks, or exhaust components, are also available in this easy-to-use medium.

MASKING SUPPLIES

Although HVLP spray guns minimize the amount of overspray in spraying primer or paint, you still need to mask off areas where you don't want extra primer and paint to end up. Even though proper masking takes a fair amount of time, it takes less time than cleaning overspray from unwanted areas.

Masking Tape

Most everyone is familiar with ordinary hardware or household-grade masking tape, but it should not be used when spraying automotive primers and paint

While spray paint is available in most every color imaginable from a variety of manufacturers, some companies also offer vehicle-specific matching colors, just like the touchup paint that comes in bottle with a brush. However, most spray can paint is not vehicle specific or designed to match any specific vehicle color code. So even if a color looks very close, realize that unless it is labeled as a specific vehicle color code, it usually isn't going to be a perfect match.

Automotive-grade paint masking tape is available in a variety of widths, with the most common being 2-inch, 1½-inch, and ¾-inch-wide (as shown from left to right). Three-quarter inch is the most common and will serve most of your masking needs. Wider sizes are especially good for masking off trim and window moldings. Only automotive-grade tape should be used when doing this type of work on your vehicle as common household tape will not endure the chemicals in automotive paint.

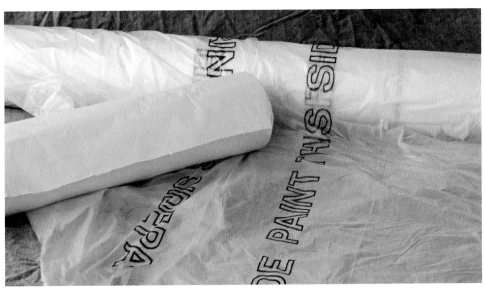

For masking areas wider than about 3 inches, use masking paper instead. It is commonly available in 12- and 18-inch widths at auto parts stores, but it is available in several different widths at your auto paint and body supplier. Masking paper can be cut easily with scissors and can be folded to fit within the desired area. If you are covering a large area to prevent it from being covered with overspray, rolls of very thin film are available. These come in long rolls about 32 inches wide, but fold out to about 10 feet wide.

automotive-grade masking tape over a headlight than having to maneuver a sheet of masking paper over that same relatively small area.

Fine Line Tape

For masking along trim or moldings that cannot be removed prior to priming or painting, use ⅛-inch 3M Fine Line tape. It is easy to use as a primary masking edge along trim and molding edges, as it is very maneuverable and will adhere securely around curves without bending or folding. After placing the fine line tape at the edge of whatever is being masked, ¾-inch masking tape can be attached to the fine line tape without the need to be right at the edge of the masked area.

Masking Paper

Rolls of quality automotive paint masking paper are available at auto body paint supply stores in widths ranging from 4 inches up to 3 feet. Masking paper is chemically treated to prevent paint or solvent from penetrating through it. Seldom will you find professional auto painters using anything but treated masking paper for any masking job. Although newspaper material may seem inexpensive and appropriate for paint masking chores, it is porous and can let paint seep through to mar surface finishes underneath. Everyone who uses masking paper will tell you that a masking paper dispenser is worth the extra money as it makes masking the vehicle much easier and faster.

Chapter 3
Repairing Rusty Sheet Metal

ICE PICK TEST

While determining that your vehicle has some rust issues is pretty easy, determining the extent of that rust can be a challenge. As you might imagine, early detection is the best, along with preventative maintenance. Sometimes rust sneaks up on us, however, so some testing is required. If you suspect that rust might be beneath your vehicle's painted surface, a simple ice pick test can reveal a lot to you. Using an ice pick, a carpenter's scratch awl, or, if necessary, a small Phillips screwdriver, gently try to poke a hole in the area that you suspect is rusty. Now don't use a hammer to drive the ice pick or similar instrument into the vehicle's sheet metal. Instead, gently probe into the suspect area. Regardless of your results, continue probing in an outward spiral pattern.

If the probe doesn't go into the metal, that is a good thing, as any discoloration is most likely just surface rust. This can be repaired easily.

If with the slightest effort, the probe goes through the sheet metal, there is indeed rust below. This detection process now becomes an evaluation process to determine how big the affected area is, which will determine the repair strategy. Continue probing outward in all directions and you will eventually find an area of sheet metal that is still solid. Using a piece of chalk, a permanent marker, or some similar marking tool, outline the affected area.

RUST REMOVAL

Multiple companies now offer more tools, materials, and supplies that can be used by do-it-yourselfers for rust removal. For years, commercial companies were commonly available in industrial areas to do this time-consuming work. However, as labor rates have increased and concern about environmental and health issues are greater, these commercial industries are not as common as they used to be. However, the processes are still basically the same, albeit on a smaller scale for the backyard enthusiast.

Chemical Dipping

Chemical dipping is not something that the hobbyist is going to pursue at home. However, the process is worth mentioning when discussing automotive rust removal, as it does work and is relatively economical depending on the size of the project. If you have numerous parts or large pieces (including complete body shells) to be stripped, having them dipped commercially is the practical way to go. The pieces and parts should be disassembled as completely as possible and any large amounts of body filler removed for the best results. Large parts will be dipped individually while small parts will be placed into a basket and then dipped.

The parts being stripped will first go into a "hot tank" that is filled with a caustic solution. This will remove wax, grease, and paint from the metal. This step will require four to eight hours, depending on the amount of buildup on the metal. When the wax, grease, and paint is gone, the metal is removed from the tank and rinsed with plain water for three to four hours to remove all of the caustic solution.

The next step places the metal parts into a second vat that is filled with de-rusting solution. While in this solution to remove rust, the material is connected to an electrical charge. Unlike chrome plating and powder coating, which use an electric charge to draw chrome or powder material to the metal, the de-rusting process reverses the current. Iron oxide molecules (rust) are drawn away from the metal and separate themselves from the good metal. Depending on the condition of the metal and the amount of rust, this step may take twenty to forty hours. When the remaining metal is removed from the de-rusting vat, the parts are thoroughly rinsed again with plain water to neutralize any remaining de-rusting solution.

Parts that are chemically stripped should be primed with epoxy primer as soon as practical to prevent the formation of surface rust. A benefit or drawback of chemical dipping is that the stripping solution will get to all surfaces of the parts that are immersed. This will remove all rust, but it may leave areas of good metal with no protection if you cannot access them to apply epoxy primer and paint or undercoating.

REMOVING RUST WITH RUST DISSOLVER GEL

Rust dissolver gel, a.k.a. naval jelly, has been around for a long time and is used quite commonly in shipyards or anywhere ocean-going boats or ships are found. However, it can certainly be used on automobiles. It works very much like paint stripper, but is typically found in smaller size containers, making it somewhat more convenient for the occasional user.

To use rust dissolver gel, wire brush any loose or flaky rust off the part to be cleaned. Then pour the gel onto the part, and brush it into all of the nooks and crannies of the part, using a disposable paintbrush. Apply it liberally. After five to ten minutes, rinse the part completely with fresh water to neutralize the solution. Most likely, some additional sanding will be required to completely clean the metal surface. This can be done with sandpaper or a wire wheel.

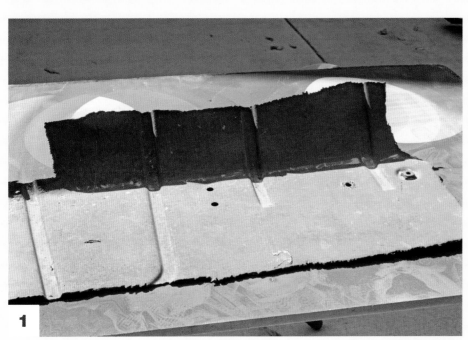

The part shown is a scrap piece of the floorboard from a 1955 Chevrolet pickup that has been replaced. Rust was not extensive, but the affected area would have been too large to sand by hand.

To provide a bit of comparison, rust dissolver gel was applied to the right half of this scrap piece. It is easy to apply with a disposable paintbrush, which allows the gel to be worked into tight areas. This gel works as a chemical reaction, so you should avoid getting it on your skin and must keep it out of your eyes. Rubber gloves and eye goggles should be considered for use when using rust dissolver gel.

As the gel begins working, it will begin changing colors, as witnessed by the pink outline near the middle of the workpiece.

3

Be patient and allow the gel to do its work. If rust is heavy, it may a good idea to apply more gel and/or leave it on longer.

4

Whenever the rust dissolver gel appears to have quit working or the part looks clean enough, flush the entire part with running water to neutralize the part. When flushed completely, all of the gel will be gone. If any gel-type substance remains, continue flushing with water.

5

Media Blasting

For a long time, course river sand was used extensively for media blasting, hence the now outdated term of sandblasting. Sand can be and is still used for removing graffiti from steel bridges, masonry structures, and various other applications. However, sand can quickly put heavy scratches into something lighter and will produce significant heat, causing sheetmetal panels to warp. Silica sand is much finer than river sand, but it has drawbacks as well. Using any kind of sand for blasting purposes will create an extremely fine dust, which is known to cause silicosis. This extremely nasty side effect can be avoided by using any of a number of alternative products.

In lieu of sandblasting, a great way to remove paint, peeling chrome, and rust is by media blasting. However, you cannot simply start blasting away at your parts.

Three very important things to remember when media blasting are:

1. Mask the area that should not be blasted.
2. Use the appropriate blasting media.
3. Remove all of the blasting media when the job is complete. You must also wear proper safety apparel, including eye and respiratory protection.

No matter what type of media is being used, media blasting will leave a slightly textured surface. For this reason, machined surfaces, bearing surfaces, threaded areas, or any other areas that would be negatively affected by this should be masked from media blasting. Exterior threads such as those on the back of some trim pieces can be easily protected from blasting by covering them with a length of appropriately sized rubber hose

Media blasting will generally leave a bit of texture to the treated surface, while chemically stripping usually doesn't. The media-blasted surface will require some additional sanding with 80–100 grit sandpaper prior to applying primer. A surface that has been chemically stripped will simply need to be cleaned really well with wax and grease remover prior to being primed.

This truck cab has received some damage to the left lower corner in its past, along with some shoddy repair thereof. The damage was evident even before the body was stripped, but it is highly evident now that some of the body filler and the primer is gone. Not that you are going to go to this extreme for any vehicle that you are simply removing rust from, but it is good to know what lies beneath the surface if you plan to do additional work.

or tubing. Other areas can be masked with heavy cardboard and masking tape, accompanied by prudent use of lower blasting pressure and a careful aim.

Media used for blasting needs to be compatible with the material upon which it is being used. If the blasting media is harder than the surface being prepped, you will do more harm than good by hurling hard objects at it. A large volume of softer material passing by the surface is a more appropriate way of freeing the surface of unwanted material such as paint or rust. These softer materials typically include silica sand, aluminum oxide, plastic media, or walnut shells. You should avoid using steel shot media or course river sand.

No one material is the best for all stripping or cleanup operations, so be sure to match the media with the task. For removing paint and rust from steel, aluminum oxide is a good choice. A little more

expensive is plastic media, which is best for stripping paint from metal, as it doesn't get as hot and cause warping. Although it is not the best media for any particular cleanup project, glass bead blasting does a good job on most any surface. For cleaning soft metals such as aluminum, die-cast, or brass, aluminum shot is best, although it is more expensive.

When choosing a blasting media, you must remember that any scratches or abrasions that you put into the metal while cleaning it will also need to be taken out. An aggressive media will no doubt remove paint and other finishes faster, but if the media is harder than the material that is being blasted, you will get to a point where you are creating more work for yourself.

The chart on the following page gives media and air pressure settings recommendations for various blasting projects.

In addition to the flaking rust, paint, or whatever was on the part prior to blasting, you must also remove all of the blasting media when the blasting is finished. Much of the media used for blasting is recycled and used again, so even if your parts were not oily or greasy, previously blasted parts may have been. Any oil that is present on your parts will cause adhesion problems, so it is imperative that all parts be cleaned thoroughly after being media blasted.

A drawback to media blasting is that the media can be difficult to remove from confined areas. It is fine for use on a simple two-sided surface such as a fender. However, on a door shell, pickup truck cab, or passenger car body, some of the media will collect between panels and will be difficult to remove. If the media would stay in hiding, it might not present much of a problem, but it will most likely decide to come out when you are in the middle of spraying the perfect top coat of paint. Years ago, I had the cab of a 1951 Chevrolet pickup sandblasted. In those particular trucks, the roof area is double-wall construction, but the back panel is single wall. Long story made short, lots of very fine sand found its way between the roof panels. Since I never got around to finishing the interior, nothing kept all of that sand in the roof, with a light sprinkling occurring most every time I hit a bump in the road.

A relatively new process that is similar to media blasting is soda blasting. In this case, the media being blasted is sodium bicarbonate (commonly known as baking soda). While this process has been shown to provide good results for removing paint during auto restoration projects, the sodium bicarbonate does not do an effective job in removing rust. So before you spend bunches of money on purchasing or renting equipment to remove paint and rust, realize that media-blasting equipment with the correct media will remove both paint and rust, while soda blasting will remove paint only.

Chemical Stripping

Chemical removal of paint and rust can be done at home or can be done commercially. When it is done commercially, it is typically the dipping process as discussed previously. The amount of stripping to be done and the availability of a chemical stripper in your area will most likely be the items to consider when this type of work is called for. If you have several parts or a number of large pieces that need to be stripped, it will be more practical to have them stripped commercially. If you simply need to strip one or two pieces or just one salvage yard fender, you can do this yourself.

Chemical stripping does not work well on thick plastic body filler, so if you have a panel that you know contains body filler, you should remove as much of that filler before attempting to strip it chemically. A simple method of doing this is with a 36-grit disc on an orbital sander. If thick body filler is not removed before the panel is chemically stripped, the filler will begin to peel, but it will not come off completely. The body filler will prevent the stripper from actually stripping the surface beneath the filler, making the entire process a waste of time.

Regardless of brand, paint stripper in general is some pretty nasty stuff. So, be sure to follow all safety precautions on the label of the product. As someone who has stripped a complete pickup truck by hand with paint stripper, I strongly recommend that you wear rubber gloves, a respirator, thick shoes or boots, long pants, and a long-sleeved shirt. If you drop a glob of paint stripper on some bare skin, you will quickly wish that you hadn't. Fortunately, it only burns for a while. At the absolute minimum, cover as much of your skin as possible and wear long rubber gloves.

BLAST MEDIA	SURFACE MATERIAL	TYPE OF BLASTING	RECOMMENDED AIR PRESSURE (PSI)
Glass bead	Aluminum, brass, die-cast	Cleaning	60
Aluminum oxide	Steel	Removal of rust and paint. To increase adhesion of paint or powder coating.	80–90
Silicon carbide	Steel	Preparation for welding	80–90
Walnut shells	Engine/transmission assemblies	Cleaning	80
Plastic media	Sheet metal	Paint removal	30–90
Aluminum shot	Aluminum, brass, die-cast	Cleaning	80–90

The one thing that the instructions for paint stripper do not tell you is that you should scuff the painted surface with 36- or 50-grit sandpaper prior to applying paint stripper. If you are dealing with decently applied paint, it becomes a standoff between the paint and the stripper if the surface is not scuffed to ease penetration of the stripper. Scuffing the surface first will allow you to use less stripper and save time in the actual stripping process.

When using paint stripper, the vehicle or parts should be on a concrete surface that can be hosed down with water after the stripping process. Check with your source of paint stripper for recommendations on disposal of the used stripper that will be rinsed from

CHEMICALLY STRIPPING SMALL PARTS

For the most part, this book is aimed toward an audience that is repairing smaller areas of rust. However, if you are working on a vehicle that is going to require any paint work and you are unsure of the history of the vehicle, it will often benefit you to get to the bare metal for the repair to work as anticipated. If you bought the vehicle as new or otherwise know that there is no plastic body filler hiding beneath, you can simply remove the rust and continue onward as described in the upcoming procedures, depending on how bad the rust damage is.

With that said, if you know that plastic body filler lies beneath, but that it has been applied correctly and has the correct contour, you are better off to avoid stripper as it will damage the filler.

You most likely will not be stripping an area the size of a hood, but a smaller part will be stripped in the same manner. Use brushes and containers that can be disposed of rather than cleaned, and be sure to protect your skin from globs of stripper that will burn for a while.

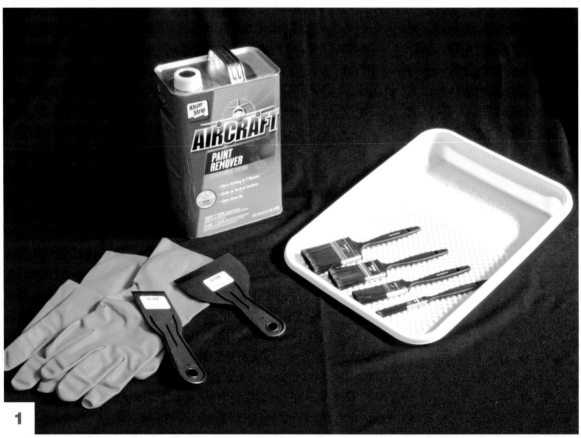

1 When using paint stripper, you will need a pair of rubber gloves to protect your hands, some paintbrushes, a small container of some sort, and a putty knife or similar scraping tool. All of these items can be found in the paint department at your local hardware store. For automotive, quality paint stripper, you will most likely need to go to an auto body paint and supply store or an auto parts store.

the parts being stripped. After scuffing the painted surface, spread the paint stripper on with a paintbrush. And then wait while the stripper works its way through the multiple layers of paint and primer. As the paint begins to bubble up, use a metal scraper or razor blade to peel the softened paint from the surface. Additional paint stripper can be applied to stubborn spots. When

all of the paint is removed from the panel, the stripper must be neutralized before proceeding with bodywork or applications of body filler or primer. Unless the particular stripper you are using calls for a different method of neutralization, use lots of water. However, you must contain the runoff and dispose of it properly and according to local codes.

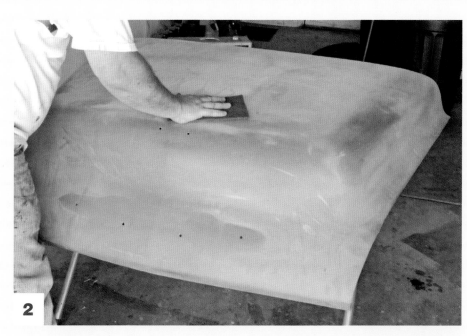

After removing the hood ornament, I scuffed the hood with a gray Scotch-Brite pad to loosen the surface and allow for better penetration of the paint stripper. We can see that the gray primer is very thin as it is coming off already, but the turquoise paint below will prove to be another story.

2

I used some disposable paint roller tray liners to pour the paint stripper into so that it could be brushed on. Most any type of flat container could be used. Just make sure that you dispose of it when the task is completed.

3

After brushing the paint stripper onto the surface, be patient and give the paint stripper enough time to work. While the Scotch-Brite pad worked well enough to loosen the gray primer, using some 100-grit sandpaper prior to applying the paint stripper would have been a good idea.

4

After five to ten minutes, the loosened paint will begin to bubble up from the surface. As the paint stripper begins to dry out, simply brush on some more.

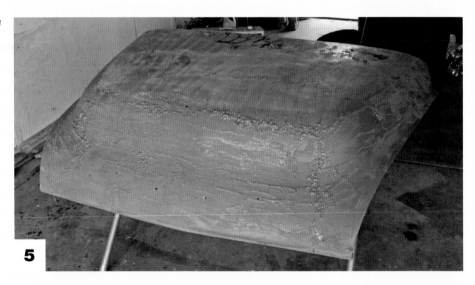

5

Although the original paint was fifty-seven years old, it is still adhering to the surface very well. After taking this photo, I scraped off the loose paint, then used a Scotch-Brite pad to scuff up the entire hood surface again.

6

7

After another, thicker application of paint stripper and waiting a bit longer, the original paint is becoming loose. It is probably safe to assume that the original paint is thicker on the front and sides of the hood because they are easier to reach and fifty-seven years of sun and washing have probably wiped away a portion of the paint from the middle of the hood.

8

I applied more stripper to remaining areas of paint with mixed results until I finally decided that all of the original paint was not going to come off this way. When finished using the paint stripper, you should remove any residue so the effects of the paint stripper are neutralized. This is done by flushing the surface completely, inside and outside, with water.

9

All of this paint should be removed prior to applying any paint or primer substrates. However, the main purpose of this exercise was to uncover any plastic body filler if any and, yes, there is some. It will need to be removed completely by using a grinder, then the imperfections in the sheet metal will be reworked. Once body filler has been covered with paint stripper, the body filler must all be removed as it is no longer a stable surface for applying paint products.

An orbital sander with 100-grit sandpaper will take off the remaining paint and make short order of removing the plastic body filler. Using a grinder to remove all of the paint would have created a lot of heat and would have caused the sheet metal to warp.

10

The white area in this photo is a thin layer of plastic body filler that will still need to be removed. The black line appears to be outlining the area that was worked prior to applying filler. The area inside the black line would have been primed and painted, while the area outside probably received some color (but no primer) while making this repair.

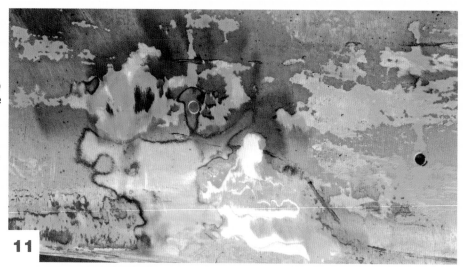

11

After just a few minutes with an orbital sander, most of the paint is gone.

12

Although I would not attempt to strip a complete car or truck by hand ever again, any one part is considerably smaller and therefore less involved. If you have the need to remove paint from one fender or a smaller panel, a gallon of paint stripper is an effective method for doing so for the do-it-yourselfer.

Sanding

Even though it is simple in theory, the sanding process of an auto body repair or paint job is probably where most problems begin. All too often, someone will grab a power sander or grinder to remove paint or rust, without giving any thought to the potential problems that may cause. Sure, the power sander will make the paint or rust removal process much easier and faster, but it may ruin a panel in the process. Any time that abrasive material is being rubbed against sheet metal, it is creating heat, even if you are doing it by hand. This heat then causes the sheet metal to warp and distort. The more heat, the more distortion. For this reason alone, extreme care must be taken when sanding, grinding, or media blasting any sheetmetal parts.

Additionally, using the wrong grit will cause problems of its own. Whether using a power sander or by hand, if the abrasive that you are using is too course for the task at hand, you will scratch the sheet metal. So when you start sanding, begin with the grit that you think is required and sand in an inconspicuous area. Is your selected abrasive removing the material that you want to remove? Is it putting scratches in the sheet metal?

If the answer to both of these questions is yes, then you should try a slightly finer grit. If the correct amount of material is being removed, without scratching, keep on sanding. It should be somewhat obvious that you do not desire to put scratches into the sheet metal, as those scratches will ultimately need to be sanded out prior to paint. However, if you are not removing the material that you need to remove, you are wasting your time. If the abrasive is not scratching, but not removing the desired material, you should try a more course grit.

Many inexperienced body repairers attempt to sand very vigorously or in a straight line. The results of this practice are sad and unnecessary. If you are sanding vigorously, you will quickly wear yourself out and you may be using an abrasive that is too fine for the task. Slow and steady is the order of the day whenever you are sanding on an automobile, whether you are taking off rust, paint, or body filler. Let the sandpaper do the work.

If you are sanding in a straight line, you will quickly sand a gouge or low spot in the sheet metal. Sanding in an X pattern and in as many different directions as possible will do a better job of removing rust or body filler, and it will also avoid creating low spots.

If you are using a power sander, keep the sanding disc moving, and support the sander so that the sanding disc is not supporting the weight. Be sure to keep the center of the sanding disc away from the sheet metal so that you do not make circular gouges in it. You should also wear eye protection and keep long hair or jewelry pulled back when using any sort off rotating equipment.

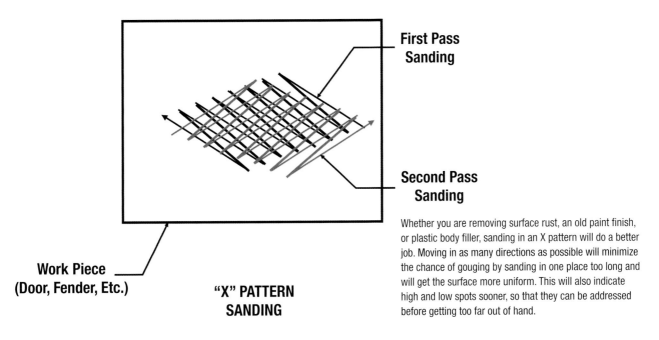

First Pass Sanding

Second Pass Sanding

Work Piece (Door, Fender, Etc.)

"X" PATTERN SANDING

Whether you are removing surface rust, an old paint finish, or plastic body filler, sanding in an X pattern will do a better job. Moving in as many directions as possible will minimize the chance of gouging by sanding in one place too long and will get the surface more uniform. This will also indicate high and low spots sooner, so that they can be addressed before getting too far out of hand.

This Subaru sedan suffers from various forms of rust or potential rust, typical of many cars today. The hood and front fenders have suffered a fair amount of road rash—gravel chips, bug chips, and anything else that it may have encountered as a part-time rally car. It was also involved in at least one collision, damaging the left rear quarter, that was not repaired as correctly as it could have been. Follow along as we sand away some of the surface rust.

The leading edge of the hood and the passenger side fender show the most signs of surface rust on this Subaru. There doesn't appear to be any bubbled rust or rust through in these panels, so they should be the easiest to repair. The basic repair strategy will be to sand down the rust, prime, and paint.

1

Perhaps a better shot of the hood and fender. The front edge of the hood appears to have been repainted at some point, with the clear coat not holding up very well. This is most likely from poor surface preparation, incompatible base coat/clear coat products, or severe road rash. The passenger fender appears to have primer coming through the paint, which could be something being spilled on the panel that ate through the paint or more road rash.

2

REPAIRING RUSTY SHEET METAL

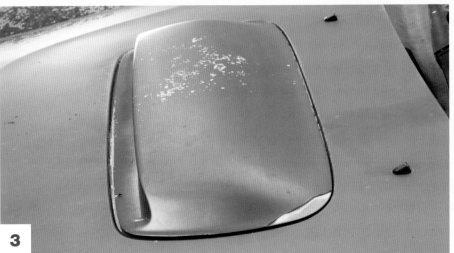

The hood scoop is aluminum, so it is not going to rust, yet it looks bad. Knowing or determining more about the vehicle in question will provide more information as to what should be done to prevent further damage. It is beginning to look like the vehicle has been painted multiple times—whether that was because of collision damage or something else is unknown.

3

Standard size sandpaper can typically be cut or torn into four strips for use on a hand-held sanding block. These two blocks take up approximately half of the sheet of sandpaper. Be sure to orient the sanding blocks on the sandpaper properly before cutting or tearing it, just to ensure that your sanding blocks are not some abnormal size.

4

Fold the sandpaper in half and then fold it in half again.

5

Using a metal straightedge placed along the folded edge, hold the straightedge down firmly with one hand and gently pull the sandpaper up to tear it with the other hand. This is really quite simple, but the amateur will often do this incorrectly, ending up with scraps of sandpaper that are not big enough to use.

6

If you were to simply tear the sandpaper, you would no doubt have non-uniform edges. Having a nice straightedge helps to prevent the edge from getting caught on something and tearing the sandpaper. While that is nothing criminal, tears in the sandpaper will make them useless and wasteful.

7

Attach the sandpaper to a sanding block to provide uniform pressure on the entire sandpaper surface. Glide the sandpaper over the area you are working on briefly and verify that you are using the correct grit abrasive. The sandpaper is scratching the paint as it cuts through, but does not appear to be scratching the bare metal surface. As suspected, the 80-grit seems to be suitable for removing the surface rust and paint from the hood.

8

The opposite side of the hood did not have primer showing, so the paint is somewhat thicker there. Additional sanding will be required, but it seems as though the appropriate grit is being used.

The passenger side seems to be cleaning up really nicely. Remember to sand in varying directions, work the entire area, and use the appropriate abrasive for the task. This particular vehicle is going through an entire repaint, so the entire hood will be sanded smooth and repainted by others. For an inexpensive surface rust repair, it is not necessary to take the entire panel down to bare metal. Removing all of the rust is really all that is necessary.

As the material being removed begins building up on the surface, use an air hose or something to blow the dust away. If the material builds up, it will begin to cake in your sandpaper, minimizing the sandpaper's effectiveness.

In the foreground is evidence of X-pattern sanding (up and down, and left and right in the photo). This part of the hood is flatter and larger, so this is easier to accomplish. Doing so does provide for a more uniform surface finish.

12

Watching your sanding process will tell you something about the general condition of the panel you are working on. These three blue spots in the sanded area indicate low spots, relative to the surrounding area. Whenever the panel in question is being worked on off the vehicle, you never really know if a panel is high or low.

13

Avoid sanding directly on body lines as you can quickly eliminate them if you are not careful. The diagonal line is a bodyline that runs from near the headlight to a point near the back edge of the hood. Since it sticks up higher than the surrounding area, primer and paint quickly comes off. The bare metal and gray primer within the narrow blue band indicate that portion might not be as uniform as could be.

14

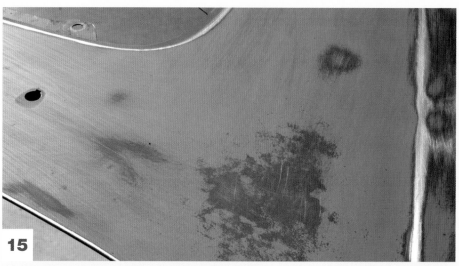

Admittedly, I'm not quite sure why primer is showing up so quickly and randomly in this relatively flat area of sanding. An educated guess would be that both base coat and clear coat from a previous repair were thin in this area. We know from other indications that the current paint is not the original factory paint job on this vehicle, so anything would be speculation.

15

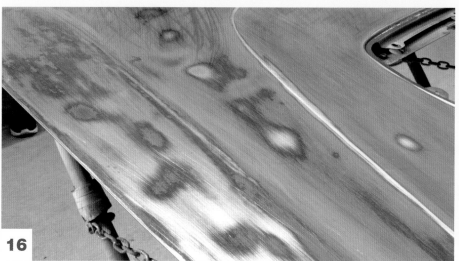

After more sanding, the low spots mentioned earlier are still prevalent. Although it has nothing to actually do with rust repair, it would be a good practice to fill these low spots prior to repainting the hood. They are not deep, but they will be noticeable when primer is applied. If left unfilled, they will be quite obvious once a shiny coat of paint is applied. Alas, we are not building show cars for this book. The choice to fill or not is up to you.

16

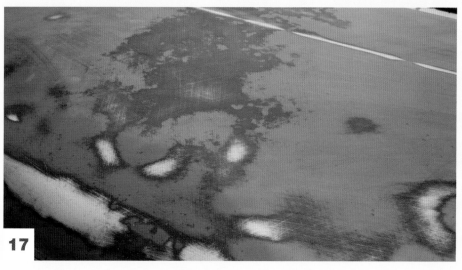

The thin paint issue continues to rear its ugly head on this hood. Without knowing the detailed history of this specific vehicle, it is purely speculation, but I would guess inadequate surface preparation prior to painting. That could be a contaminated surface, poor mixing of either base coat or clear coat, or some sort of compatibility issue between paint components.

17

REPAIRING RUSTY SHEET METAL

More high spots and low spots, which are decreasing the credibility of the previous paint job. If you are simply doing rust repair, getting rid of the rust, spraying some primer, and some paint of a reasonably close color are an improvement. You need only to please yourself.

18

The leading edge of most any vehicle's hood is going to be subjected to rock chips and bug splatters, making it susceptible to the formation of surface rust. If it is not resolved as surface rust, the damage will continue to spread and worsen. Applying touchup paint as the initial damage occurs is significantly easier and less involved as sanding the surface, priming, and painting.

19

The surface rust has been sanded away, and the rough spots sanded smooth. If so inclined, now would be the time to use plastic body to fill any low spots or perform any other bodywork that you became aware of during this process.

20

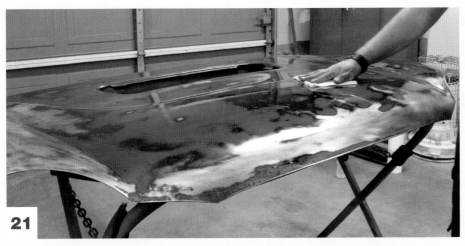

Quality bodywork and painting requires that the surface to be painted be as clean as possible. Use an air hose or other source of compressed air, to blow away any dust, dirt, or rust particles. Then spray some wax and grease remover onto a clean paper towel or directly onto the surface. Wipe the entire surface and then look at the contact side of the paper towel. Most likely, you will be surprised at how much stuff it picks up.

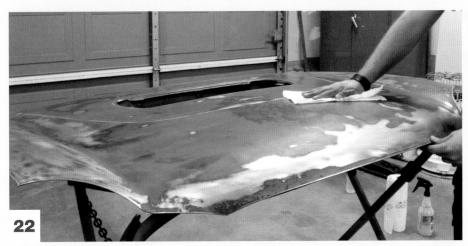

Using another clean paper towel, dry the surface. Do not allow the wax and grease remover to dry on the surface.

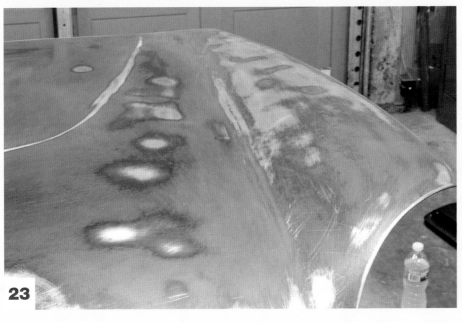

The hood is now ready for a coat of filler primer. If a professional body shop was doing this repair, they would most likely do a bit more sanding with finer abrasive, such as 180-grit. They would then spray epoxy primer onto the bare metal spots, if not the entire hood, as they would have ready access to an air compressor and spray gun. After applying epoxy primer, a body shop would fill the low spots with filler, then sand the entire surface with 220- or 240-grit sandpaper. Additional primer would then be sanded with 400-grit sandpaper, prior to being painted.

BASIC BODYWORK

Most of this chapter is going to made up of photo sequences showing actual rust damage repairs being made to various vehicles. As the damage gets worse, more work is required, which is to be expected. Some of the basic procedures are repeated in some, if not all, of this work. However, if you are reading this book, you might not be familiar with some of those basic bodywork steps. Rather than attempt to explain those basic steps while trying to explain rust repair, we'll go ahead and explain some of the basic steps now, so that you can learn those first, and then concentrate on repairing rust a little later.

Hammering on Sheet Metal

The first basic step is hammering sheet metal. You will be using this not only to straighten out dents, but to create small patch panels if you have areas of rust where

the original metal is completely gone. In most situations, using a hammer by itself (without a dolly behind the metal) will merely cause a larger-than-desired area of metal to move inward. This is usually not what is needed to repair a dent in a piece of sheet metal. By using a dolly behind the panel and a hammer in front of the panel, the sheet metal can be worked in a more predictable manner as the dolly focuses the force of the hammer.

Dollies are used in two basic ways: dolly on or dolly off. When hammering on the dolly, the dolly is located behind the sheet metal and directly beneath the hammer blow. This method is used to knock down high spots or to smooth ripples within the relatively small size of the dolly. Hammering off the dolly is done by hitting the surface of the panel adjacent to the dolly, rather than on it. This causes the dolly to push outward while the hammer pushes inward and is typically used on larger areas of repair.

This exaggerated drawing should explain the differences between dolly-off and dolly-on hammering. Quite simply, it is the relationship between the position of the body hammer and the body dolly when striking a piece of sheet metal.

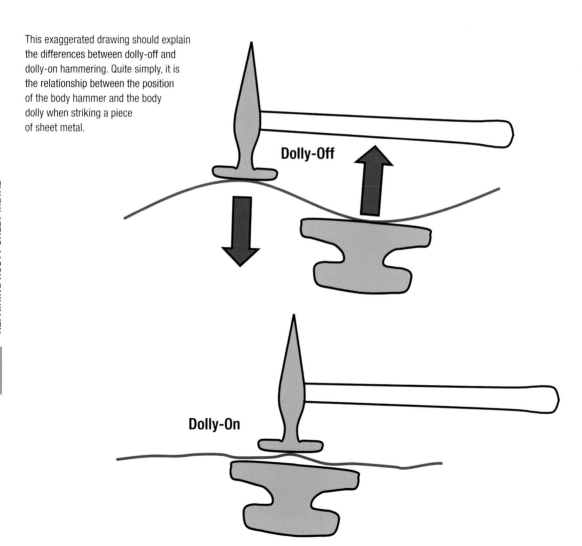

Dolly-Off

Dolly-On

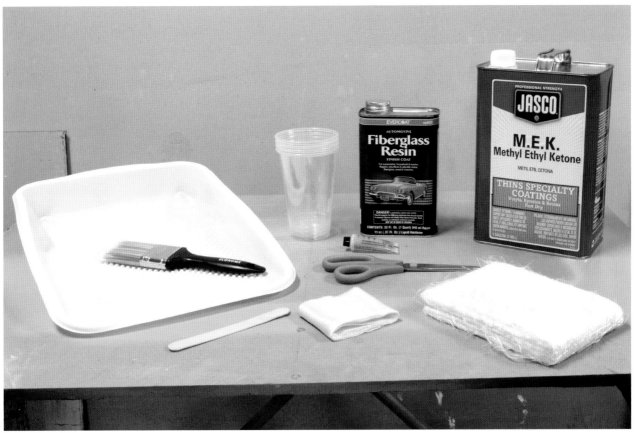

When working with fiberglass, some basic tools and materials are required. From the left, some sort of pan to mix the resin and hardener in; a cheap, disposable brush for applying the resin to the cloth or mat material; mixing cups for mixing small amounts of resin; resin and hardener; MEK (methyl ethyl ketone) or acetone for cleanup; fiberglass mat; scissors; fiberglass cloth; and a tongue depressor or similar for stirring the resin and hardener.

REPLACING SHEET METAL

Whether the sheet metal is bubbled or rusted through completely, it needs to be replaced. If it is bubbled, it is flaky like a biscuit, so there is no way to put it back together. The remaining paint on the surface is all that is holding it together. If it is rusted through, the sheet metal is already gone. There are two basic choices if you intend to repair either one of these situations. One is to replace with fiberglass, the other is to replace with sheet metal. A professional body shop would cut out the rusty area and then weld in new sheet metal, but that is beyond the scope of this book.

For a budget rust repair on a daily driver, either material is acceptable. Whichever one you feel more comfortable with is the best choice for you and your repair.

Using Fiberglass

Given a little bit of practice, fiberglass is relatively easy to work with. Of course, my expert metalworking friends would disagree. I suppose fiberglass is easier as it is not as precise. Fiberglass mat or cloth is saturated with fiberglass resin that has been mixed with a catalyst. Whatever shape that fiberglass material is in when the resin cures is the shape that it will remain. For this reason, fiberglass is great for patching a small hole or area that provides adequate support.

If a hole is on the lower side of a panel, it may be worthwhile to remove the panel from the vehicle and complete the repair on a workbench or workstand. This would allow gravity to hold the fiberglass in place while it cures, rather than pulling it away from a vertical panel.

If a larger panel or area requires replacement, thin cardboard could be used to temporarily support the fiberglass material and resin during the curing process. The cardboard would then be removed from the inside (or backside) afterward.

Whenever you are working with fiberglass, set up some sort of workbench, even if it is temporary.

Working on the garage floor gets tiresome in a hurry. Also, make sure that you have some MEK or acetone for cleanup purposes.

Mixing

Anything that you are mixing fiberglass resin and hardener in should be disposable, but you should have some MEK or acetone to wash fiberglass resin from your hands or any other body parts that you may get it on. The first time that you drop a big glob of resin and hardener on some bare skin and it begins to cure, you will learn that this chemical reaction creates a bunch of heat.

You should also wear some type of disposable gloves and eye protection. Especially if you start sloshing the resin around. It is safe, but it is a chemical that creates lots of heat in the process, so be careful.

Prior to mixing fiberglass resin, it may be a good idea to cut out the pieces of fiberglass mat or cloth to the shape and size that you plan to use. Doing this will allow you to use scissors before you put on rubber gloves, which can sometimes be a challenge. Having the pieces or fiberglass material cut out beforehand may also ease the anxiety that a novice user may have, even though fiberglass resin does have a decent working life.

If you are going to be dipping the fiberglass mat or cloth, use a flat pan of some sort for mixing the resin and hardener. A disposable paint roller tray liner works great for this. If it is going to be more convenient to pour resin onto the part, use a disposable drink cup. After choosing your mixing container, pour in what looks like the correct amount of resin. The fiberglass resin is a thick liquid, usually brownish in color. The hardener is usually clear, but the mixture will change color as the two components are mixed together. Like many other two-part auto body repair products, add a proportional amount of the hardener to the resin. Then use a tongue depressor, paint stir stick, or other similar tool to thoroughly mix the resin and hardener. If there are streaks of color, it is not yet mixed thoroughly.

Remember that fiberglass mat is used to create stiffness, to add bulk, or in a situation that calls for complex shapes. Fiberglass cloth is thinner and contains more glass than mat, so it is more suitable for flatter shapes and for creating strength. If you are going to be using both, either in a new part or repair, begin with mat to build stiffness and then cover it with cloth to build strength.

Shaping

Whether it is fiberglass mat or cloth, dip the first piece of material into the fiberglass resin so that it is completely saturated. It can then be placed onto the sheet metal. Use a plastic body filler spreader or disposable paintbrush to press the fiberglass material in place and to force out any bubbles. It is important to eliminate any air bubbles that form.

After the first piece of material is situated as desired, lay on the next piece of fiberglass material. Since it has not been saturated, it will be a little bit easier to position. When it is in position, pour or brush resin onto the latest piece of material, making sure that it becomes fully saturated. As before, use a spreader or brush to push out any air bubbles and conform the new material to any creases or crevices. Continue this process until the fiberglass is in the desired shape. Know that you must use some plastic body filler to achieve the final desired shape and surface finish that you want. Allow the fiberglass to cure prior to applying any plastic body filler. The cure time will vary somewhat, depending on ambient temperature and humidity. If the repaired area is still giving off heat or the material is still soft, it has not cured.

Using Sheet Metal

Since the bulk of the vehicle's body is made of sheet metal, it should make sense to replace it with sheet metal … even though it did rust. If you have the necessary tools and ability to shape sheet metal, this can be an easy repair. Years ago, sheetmetal patches would be welded in place of rusted-out panels, whether it was a small repair or a complete panel replacement. That is still a valid method of rust repair.

However, the automotive manufacturing process has changed, making rust repair easier than it used to be. Most, if not all, new OEM vehicles are being built with a modular process. No longer does a car's main body consist of side panels, floor, and roof that are welded together into one unit. Instead, more of a steel framework is implemented to support the outer panels that are glued to it. This framework is lighter, because it is more of a skeleton. Sheetmetal or composite body panels are then attached to provide the outer skin. Smaller panels that are glued on make for an easier repair, whether it is due to collision or rust repair.

So how does this affect your rust repair? The answer is quite simple, even if your vehicle was not

COVERING A SIMPLE SHAPE

With the sheet metal of contemporary automobiles being significantly thinner than their twenty-plus-year-old counterparts, that sheet metal is more susceptible to impact damage. If not repaired in a timely manner, that impact damage can quickly become rust repair as well.

In the following photos is a daily drive/weekend rally import tuner car that suffered some collision damage during a race. Left unattended, the broken paint did indeed allow the sheet metal beneath to turn to rust. Basically, you can address this repair in two ways. One is to purchase a replacement fender, whether it is new, used, or aftermarket. Depending on the source, the vehicle in question, and the damage, that can be an expensive proposition, which is probably why it was not addressed before rust became involved. The second method is to repair or at least cover the damage.

Again, combining the thin gauge of sheet metal involved and the fact that it is in a narrow portion of the fender, hammer and dolly work does not seem very feasible in this instance. It would be very easy to beat the fender into oblivion unless great finesse is used. As is, the dent is way too deep to attempt to merely fill it with plastic body filler. So to get this fender looking better, use fiberglass to fill and strengthen the area, and then finish that with body filler. It may not be the perfect repair, but when finished, it will no doubt look better than it does now.

1

This daily driver/weekend rally car suffered some collision damage to the left front fender. Since the impact cracked the paint and was not repaired in a timely fashion, rust crept in. With the fender being stamped out of thin 24-gauge sheet metal, hammering it back into shape will require more finesse than anything else.

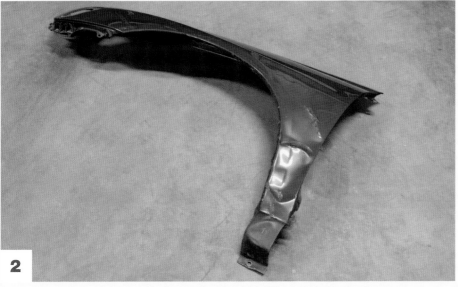

2

With that fender off the car, you can get a better look at the damage. It would be very easy to totally destroy this fender, just crumbling it up by hand, so using a hammer is out at this point. An experienced body repairer may be able to pull it off, but the cost might be more than a replacement fender.

With the fender caved in, the stock sheet metal is simply too deep to attempt to fill with plastic body filler. Doing so would add considerable weight to the fender and would most likely not hold up very well.

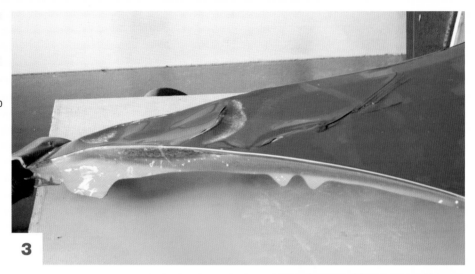

3

Before commencing any repair procedure, the surface should be cleaned with wax and grease remover, so that any contaminants that are on the sheet metal are not further engrained into the fender. Either spray the wax and grease remover onto the surface or wipe it on with a clean paper towel. Then wipe it off with another clean paper towel.

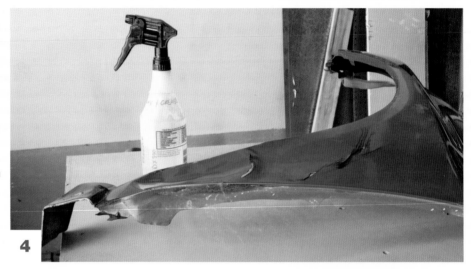

4

Now we need to do some investigating to see exactly what lies beneath the existing paint and verify that no more rust than what we could see originally exists. To do this, use a sanding disc or a wire wheel attachment on a sander or drill motor. If the surface was relatively flat, a sanding block could be used, but that method would not get into the nooks and crannies of this fender.

5

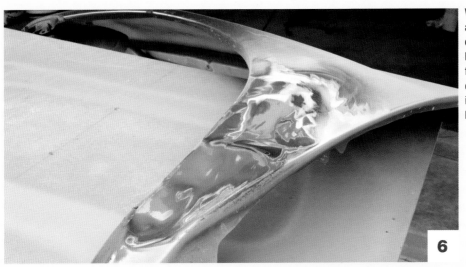

With just a bit of effort with a wire wheel, it becomes obvious that this fender has had some plastic body filler applied in the past, as evidenced by the white areas in the photo. The gray area is bare sheet metal.

6

Continuing with the wire wheel, all the paint has now been removed from the affected area and its surrounding area. The white areas are still plastic body filler, but what is there seems solid so it does not need to be removed. If this area had been stripped with chemical stripper of any kind, it would need to be removed completely as its integrity would be compromised by that stripper.

7

After using an air hose to blow away (or a paper towel to wipe away) any sanding dust, clean the surface again with wax and grease remover as described previously to remove any contaminants from the surface.

8

To help restore some strength to the damaged fender, fiberglass mat and then some fiberglass cloth are going to be glassed in. To avoid getting fiberglass resin all over the scissors, cut out the strips of fiberglass mat and cloth before mixing any resin.

The fiberglass mat and resin will be much easier to trim while it is dry, so trim it to the desired size and shape. It will need to be multiple layers at the dent's deepest area, but then feathered out as it gets back to the stock profile. Be sure to allow for at least a skim coat of plastic body filler to transition back to the original sheet metal. In other words, avoid building up too much thickness of fiberglass.

These smaller pieces will be placed as shown, atop the longer piece that will be applied first. No, it probably doesn't matter if the small pieces are applied first or last. Hindsight being 20/20, it might have been better to apply the smaller pieces to the deeper areas first and then the larger piece last to provide a more uniform surface.

When using fiberglass resin, think disposable rather than cleanable. Using an inexpensive tray, such as this household paint roller pan liner, pour in an amount of fiberglass resin, and then add in a proportionate amount of hardener (as an example, if you use a quarter of the can of fiberglass resin, use a quarter of the tube of hardener that came with it). Use a paint mixing stick, tongue depressor, or plastic spoon, or similar object to mix the two components together. As with most two-part products, mixing them together will change the color, so keep stirring until the color is even throughout.

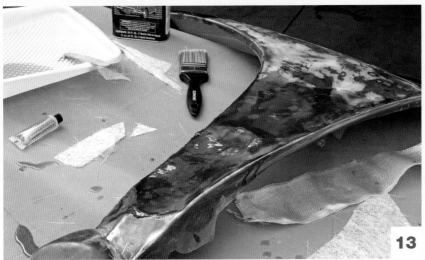

Whether you apply the smaller or larger pieces of fiberglass mat first, dip the first layer pieces into the resin so that the mat is fully saturated. Then place the saturated fiberglass mat on the fender. Using a cheap paintbrush, press the saturated mat into place so that it is not wrinkled or bunched up.

Before the fiberglass resin kicks (begins to cure), apply the additional layers of fiberglass mat, by laying them in place and then saturating them by brushing the resin on. Verify that each additional layer of mat is completely saturated with resin before adding additional layers. Use a cheap paintbrush or filler spreader to work out any bubbles or voids in the resin.

After the area is built up with fiberglass mat, add a layer of two of fiberglass cloth to provide a smoother surface. Add the cloth the same way as the mat, cut and trim it to shape, and saturate it completely with fiberglass resin.

Verify that all of the area being repaired is saturated with fiberglass resin, especially the edges of fiberglass mat or cloth.

Now, just let the fiberglass resin cure completely. Direct sunlight will speed up the curing process and fresh air will minimize the unpleasant (to some) smell from building up in your garage or workspace. After the fiberglass cures, the affected area will be ready for the application of plastic body filler to bring the fender back into the correct contours. This will be discussed in the "Applying Body Filler over a Fiberglass Repair" section.

constructed in this new manner. As you might imagine, the OEMs and their venders have developed some hard-core adhesive for attaching these panels. What is good for the automotive industry is good for the auto body repair industry, and is now good for the vehicle owner who makes their own repairs. You too can use panel bond adhesive to repair your rusted-out body panels. No more need for rivet guns for attaching sheet metal.

Before you commence to beating a piece of sheet metal into oblivion, check to see if patch panels for the area that you need to repair are available. Depending on the age, make, and model of your car or truck, this will range from amazingly easy to painfully difficult. However, having a piece of stamped sheet metal that is the same size and shape of the area you need to replace is a very good place to start. The more common rust-prone areas on any vehicle are more likely to be available. For popular vehicles, especially those that are deemed collectible, multiple size panels may be available for the same area—a small patch that covers just the top of a front wheelwell, a larger patch that covers the entire wheelwell lip, or an even larger patch that includes the area between the wheelwell and the door are an example. So, you may not even be required to purchase an entire fender. Of course, depending on the extent of the rust, an entire replacement fender may be worth the money for the time and effort saved.

If the aftermarket does support your vehicle and multiple patch panels are available for the area that you need, do yourself a favor and buy the size larger than what you think you need. You can always trim the patch panel if you have more than what is needed. Also, you must understand that sheetmetal patch panels typically are not going to be a precise fit, especially when you have multiple body lines within that same area. In order to get those body lines aligned, you may need to cut the patch panel into two pieces, and then place them to match the body lines. Then trim the middle excess if the patch had excess between the body lines. Or, add some sheet metal in between the two if it was too small. Either way, a stamped sheetmetal piece is probably going to be a better overall fit that anything that you could bend out of a piece of flat sheetmetal stock.

Okay, you have looked and cannot find the necessary patch panel for your vehicle. Have you looked for a similar part in your local salvage yard or used car parts ads in the newspaper or online? Even if you could find a damaged fender, door, or whatever you need that isn't damaged in the area that you need to replace, that would be a start. Even if this other used part is not good enough to simply install on your vehicle, it could be the source for a similar-shaped patch panel.

If you still don't have any luck, you may have to obtain a piece of sheet metal (18–22 gauge) in the appropriate size and try your luck at becoming a metalsmith. Using a smooth-faced hammer, a bit of finesse, a bit of creativity, and some patience, you can probably make a simple metal replacement. If not, fiberglass might not be so bad after all.

Making a Pattern

For patternmaking of sheetmetal projects, poster board is ideal. It is inexpensive, readily available, easy to work with, and is ultimately about the same thickness of sheet metal. Using a pen, pencil, or felt-tipped marker and a pair of scissors, you can create a pattern for the sheetmetal patch that you need. Even if the piece that you need is three dimensional, you can make a flat pattern of each general area of the patch, then tape them together into the overall shape. The more accurately that you can make the pattern, the more accurately you can make a real patch panel.

Gluing a Sheetmetal Patch in Place

With patterns and sheet metal in hand, two flat panels will be cut out and glued to the upper firewall of this truck project. The holes could have been welded shut, but with the number of them, that would have taken quite a while.

Like most any other bodywork repair process, cleanliness is key. Use wax and grease remover to clean the surface that is going to be covered up. Test-fit the panels one last time to verify that they are correct and that you are not covering up anything that you do not intend to. If you make any additional adjustments to the sheet metal, double check it for fit again.

Determine the method of clamping, using C-clamps, magnets, or whatever you have to use. There will be a fixed amount of time available to get everything in place, but once that time is up, the sheet metal will be stuck where you have it. Scuff both mating surfaces with 80-grit sandpaper. Then, while wearing rubber gloves, clean mating surfaces again with wax and grease remover. Apply panel bonding adhesive to both surfaces in accordance with the adhesive's directions. Position the sheet metal and clamp in place. Leave the clamps in place for the duration of the curing phase.

MAKING A SHEETMETAL PATTERN

On the author's 1955 Chevy pickup truck project, the upper firewall currently includes a collection of various sized holes. They are not rust holes, but they are holes that need to be eliminated, so they will serve as a great example of making a pattern for two pieces of sheet metal that will ultimately be attached to the firewall with panel bond adhesive.

The final pattern will be made out of poster board, but to determine the basic size and shape, some Kraft paper will be used. Cut out a piece of Kraft paper that will be somewhat larger than the area to be covered. In this particular case, some welding magnets will be used to hold the Kraft paper in place. This will allow you to press the paper into the edges of the existing sheetmetal body and then trace around that edge with a permanent marker.

After tracing around the entire perimeter, remove the welding magnets and lay the Kraft paper onto a piece of poster board. Use something to hold the Kraft paper in place on the poster board, and then trace around the edges of the Kraft paper with a marker. Prior to cutting out the pattern, look for any areas that should obviously be straight, and straighten those out with a straightedge. If there are curved areas that need to be smoothed out, do that at this time too. When you are satisfied with the outline of your pattern, cut it out with a pair of scissors.

Now check the pattern with the vehicle. Does any portion of the pattern need to be reshaped, cut off, or made larger? If any portion of the pattern needs to be larger, cut out additional pieces of poster board and tape them to your pattern and trim as needed. If the overall pattern is too long or too wide, find a place to cut out a section and then tape the pieces back together. If the pattern is to short or too narrow, cut the pattern in two and then add a section.

Once your pattern is the desired shape and size, mark some reference points on it (if applicable) so that it can be situated in the same location again. If any bolt holes or anything else that the pattern must allow for exists, locate those at this time.

1

Several holes in the upper firewall of this '55 Chevrolet pickup cab can relatively easily be eliminated with some sheet metal and panel bond adhesive, just as if we were covering a rust-through situation.

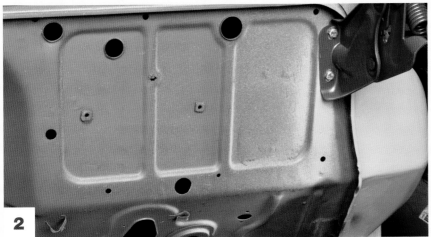

The driver's side should be the easiest of the two sides, yet it does present a couple of minor obstacles. There are two holes where the hood hinges mount on each side. These are threaded holes and the location is critical toward getting the hood to open and close properly. On this side, however, the three largest holes have a bit of a flange around them. To allow the replacement sheet metal to lie flat, these flanges will ground down to below the surface level of the mounting surface.

The passenger side also includes the stock battery tray. These are threaded holes also, which would not matter if the battery was going to be relocated, but I'll be going with this stock location. Doing this will call for an excellent pattern, to ensure that the holes in the sheet metal are located properly.

The hood hinge is secured by two bolts that go through the hinge's mounting bracket and into threaded holes in the firewall. Using a wrench of the proper size, remove the hood hinges.

By using a grinding wheel in a die grinder, grind off the flange from each of the three holes. Verify that you have the entire flange ground off.

5

Then use an 80-grit sanding wheel in an angle-head grinder to smooth up any loose edges. Be sure to wear eye protection whenever you are grinding. Mechanic's gloves are good protection when handling sheet metal.

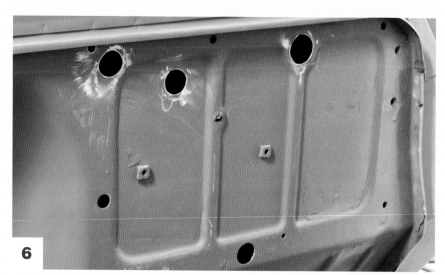

6

The area to be covered is surrounded by a flange on the top and outside. Along the bottom of the area to be covered, the firewall runs into the toe board, creating a line. The area toward the center of the firewall is the recess for the distributer and is a smooth curve. One the lower and center edges, plastic body filler will be used to blend the filler panel and the original together.

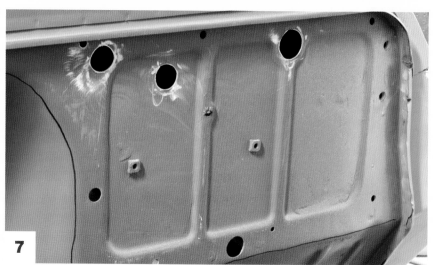

7

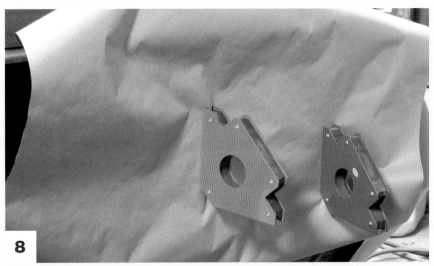

After cutting out a piece of Kraft paper large enough to cover the area in question, hold it in place with magnetic welding clamps.

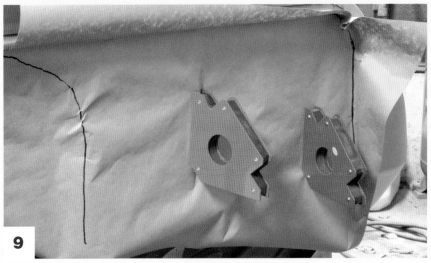

Press the Kraft paper along the crease made at the intersection of the firewall and the flange that surrounds part of it. Use a permanent marker to draw a line in this crease. Then draw a line along the center edge. This last line won't be quite as accurate, as it will be somewhat arbitrary.

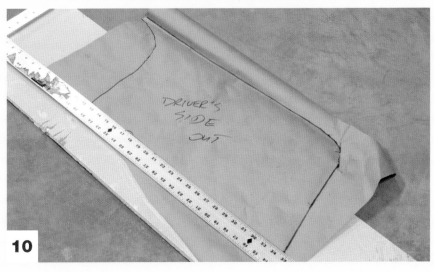

Since the bottom edge is a straight line, make a mark at each end and then use a straightedge to connect the two with a marker.

Use scissors to cut out the pattern and then check it for proper fit. Again, magnetic welding clamps can be used to hold the pattern in place. Be sure to mark the locations for any holes that should not be covered.

11

Lay the Kraft paper pattern onto a piece of poster board. Trace around the paper pattern, and then cut out the poster board.

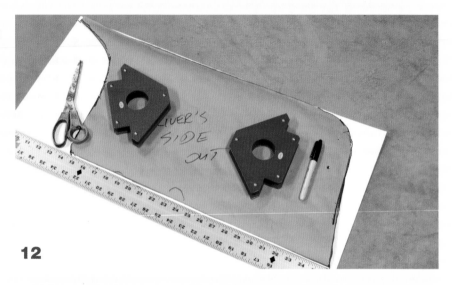

12

Fit the poster board pattern in place, trimming if necessary; secure it with magnets, and then determine the exact location of any holes that are to remain.

13

14

After locating the holes, use screwdrivers or some similar device to verify that the pattern does not move. At this point, make any modifications to the pattern that are required. In this case, some small pieces of poster board were added near the flange and taped in place with painter's masking tape.

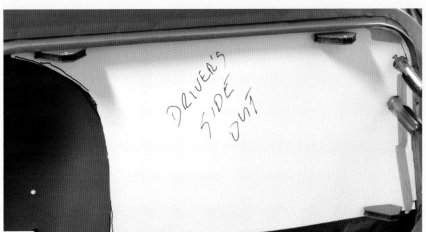

15

The next step is going to be cutting sheetmetal base on the pattern, so double check it yet again. Kraft paper and poster board is cheaper and much easier to trim or expand than sheet metal.

16

Lay the completed pattern out on the sheet metal (24-gauge in this case) and secure it in place. Again, the magnetic welding clamps work well. Trace around the pattern with a scribe or permanent marker. Note that as the permanent marker gets broader, it will give a wider line. This will provide you with a larger safety margin for error, yet may require more trimming.

After the clamps have been removed, perform the necessary bodywork to properly transition the surface of the sheetmetal patch to the surrounding original sheet metal. Then prime, mask, and paint.

Follow along as we smooth up the firewall on this Chevrolet Task Force pickup.

Doors and Door Skins

Since automobile doors consist of two basic components—a mostly hollow inner panel and an outer skin—some door repairs can be made by simply reskinning the door. However, you should consider some things before going this route. First and foremost, this repair will be beneficial only if the original inner panel is still undamaged (or at least straight). If collision impact caused damage to the inner panel that cannot be easily repaired, you should consider using a replacement door. Likewise, if the outer skin is damaged due to rust, you should verify that the inner panel is still structurally sound. It may be usable, but then again it might actually be in worse shape than the outer skin.

If the inner panel is still usable, you will need to determine if an outer skin is available for your make and model of vehicle. If it is, you are in luck as a door skin

GLUING METAL IN PLACE

Automotive panel boding adhesive from 3M comes highly recommended as the product to use for gluing a piece of sheet metal over the upper firewall of this truck project. As an epoxy, it comprises of two different products, that when combined should be more than adequate for securing a piece of flat sheet metal onto the firewall of this truck project. This is the same product that most OEM facilities use to attach panels together at the factory.

3M's Panel Bonding Adhesive is a two-part product that is packaged so that it is easy to use by the consumer. It comes with two tubes, allowing the user to use part of the product at one time and the rest later.

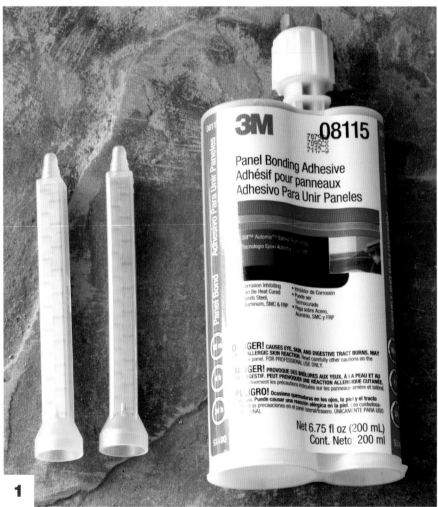

1

will be considerably less expensive than a replacement door. Of course, depending on if this is your daily driver that you are repairing or a ground-up restoration, you may choose to replace the entire door while you are at it.

If you decide to replace the door skin, it can be done as follows: To make the work easier on yourself, remove the door from the vehicle, the internal components from the inside of the door, and then set the door atop a pair of sawhorses or other suitable work stand.

The original door skin is secured by wrapping the edges around the flange of the inner door panel. To remove it, use a grinder along the edge of the door to separate the main part of the door skin from the part that flaps over. When this is completed, you should be able to remove the skin. If it cannot be removed, there may be spot welds around the flange. If there are, you will need to drill them out. You should now be able to remove the outer skin. You will also need to remove the part that flaps over from the inside of the inner door panel.

To install a new skin, you should first double check to verify that no fragments of the previous door skin, any spot welds, or panel adhesive remain on the surface to which the new skin will be applied. Then position

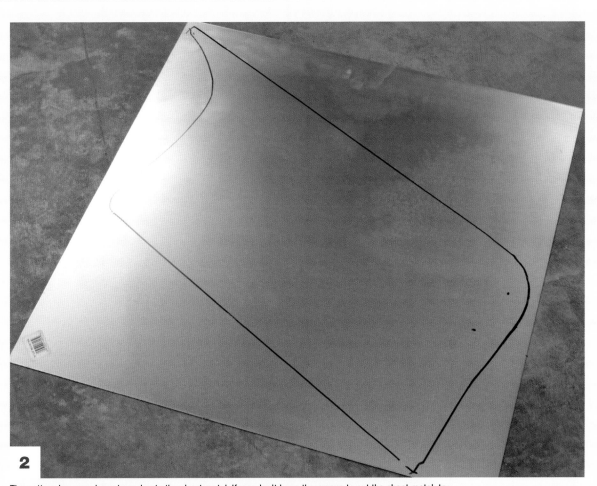

2

The pattern has now been traced onto the sheet metal. If you don't have the means to cut the sheet metal, try contacting a heating and cooling shop because they cut sheetmetal ductwork all the time. They may charge a nominal fee, but it may be worth it.

Drill the required hood hinge holes so that the mounting bolts can work to secure the sheet metal in place. I found that the welding magnets on one side of the firewall were not strong enough to hold the sheet metal in place. However, placing welding magnets on the inside of the firewall and on the outside in the same location did work.

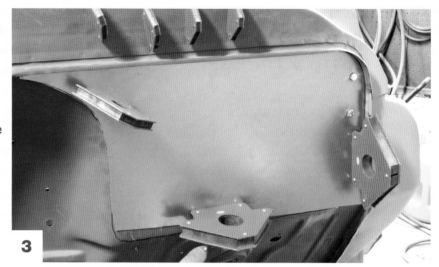

The pattern was good, but an incorrect cut was made while using throatless shears. As you can see in the photo, the bottom line on the sheet metal was cut at a bad angle. It is a tad too low on the outside and about three quarters of an inch too high at the center corner.

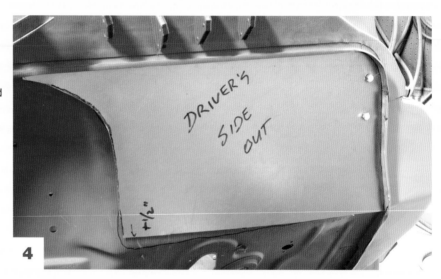

Since a second piece is going to be required, trim this new pattern as required for a perfect fit.

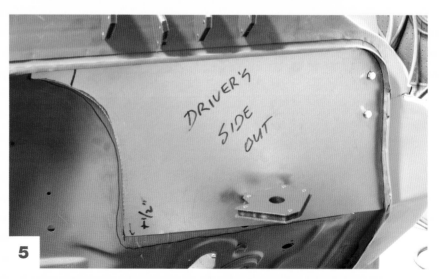

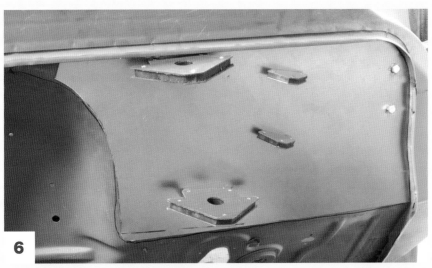

After cutting a replacement panel, check it for fit. Verify that everything is as you want it.

6

7

The marker lines indicate approximately where the ribbed surfaces of the firewall will be. This is where the firewall and the filler panel will make contact, therefore where the panel bond adhesive will be applied.

Use an 80-grit wheel on an angle-head grinder or 80-grit sandpaper to scuff up the areas on the firewall that will make contact with the filler panel. Then clean the surface with wax and grease remover.

Likewise, scuff the backside of the filler panel in the appropriate areas with 80-grit, and then clean with wax and grease remover.

Follow the instructions of the specific panel bonding adhesive product to apply that product to the area to be covered.

Apply the panel bond adhesive to the backside of the filler panel as well.

Position the filler panel and then clamp it into place. Any threaded holes that are not to be covered serve as excellent clamping points by using a large washer secured by the proper size bolt for the threaded hole.

After the panel bond adhesive has cured (refer to the instructions for the specific product that you use), remove any clamps, magnets, or other clamping devices. Finish the area as required with plastic body filler, primer, and paint.

the new skin in place so that the excess "skin" is centered front to back, and top to bottom. Using a permanent marker or a scribe, make some positioning reference marks on the inside of the door skin so that it can be placed in this position again. Then remove the door skin and apply a bead of panel adhesive to the inside of the door skin and to the mating surface on the inner door panel. Allow the panel adhesive to get slightly tacky, and then press the door skin into position. Use clamps to secure the door skin in place. You can use C-clamps, clamping pliers (Vise-Grips), or other types of clamps. However, you must take the appropriate precautions to verify that you do not damage the new door skin or inner panel. After the two panels are clamped together, wipe away any excess panel adhesive that oozes from inside the door.

REPLACING A DOOR SKIN

To see how an old door skin is removed and a replacement installed, I visited Morfab Customs where Chris was in the process of reskinning an early Camaro door. The process sounds more difficult than it actually is, and is oftentimes a much better method of repairing a damaged door.

At the time this is written, a replacement door skin for this vehicle costs about a hundred bucks, while a replacement door costs around five times that much from the same source. Since the need for this repair in this case is rust through due to some amateur bodywork in the past, a replacement door skin is the feasible way to go.

1

After the panel adhesive has set (refer to the product label), the edges of the door skin still need to be folded over the edge of the inner door panel. There are several ways to do this and even more tools available to do it. Some body repairers use door skinning pliers to fold the edge over, while others use duckbill locking pliers. After the edge is folded over somewhat, it must be pressed down flat against the inside of the door panel. Some use a light, door skinning hammer, while others use a mallet or a smoothed-face body hammer and a dolly. This is one of those cases where there are several ways to do the same task, depending on what tools you have available to you and how you use them. The main thing to remember is that the edge of the door skin must be flat against the inner panel, but you don't want to do any damage to the outer side.

2

To remove a door skin, you should use a grinder with a 36- or 80-grit disc to grind through the edge of the door skin that is to be removed. You should not grind the face of the door or the flap that is folded over, but the very edge. This will quickly separate the skin from the door.

With the door skin separated at the door's edge, the skin can be removed from the outside of the door and the flap can be removed from the inside of the inner door panel. It may be necessary to drill out spot welds from some doors to remove the skin. With the door skin separated at the edge, the skin is now ready to be removed.

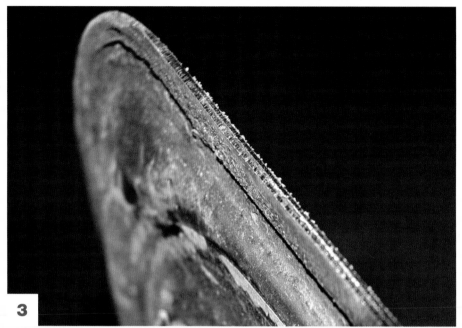

3

If you were so inclined, now would be a good time to have the inner door panel media blasted or chemically dipped to remove any rust, and then follow up with a coat of epoxy primer. This door is solid, but lots of surface rust is apparent in the photo. Between removing the skin and replacing it, is the only time that you can effectively do any surface treatment to the inside of the door.

4

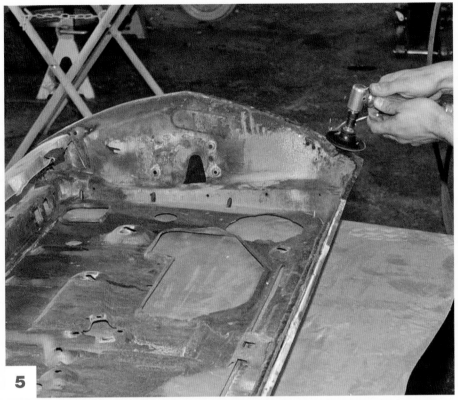

The replacement skin will be glued in place and then the edges of the skin wrapped around the inner panel, much like originally. So that the glue will form a good bond, the mating area of the inner panel is scuffed with an angle-head grinder to remove any paint or rust.

5

Likewise, the mating surface of the replacement skin is scuffed in similar fashion. The original doors did not use glue to attach the door skins, but it does make for a better repair.

6

The glue used for securing a door skin is 3M's Door Skin Adhesive. Pretty simple, huh? It is a two-part epoxy that does require a specialized application gun that mixes the two parts as they are dispensed. Since the two parts are not mixed until they are used, the product has a longer shelf life.

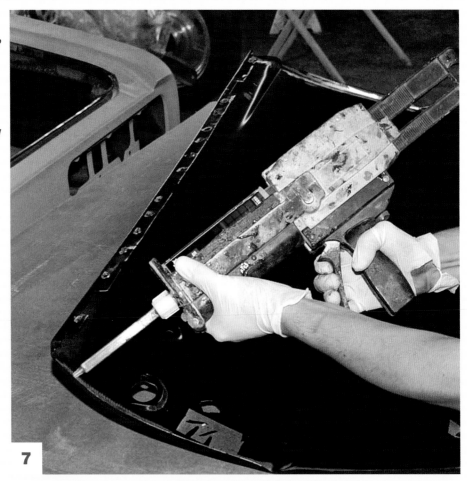

7

Used much like a caulking gun for various caulking chores around the house, Chris begins at the top of one side of the replacement door skin, working his way down that side.

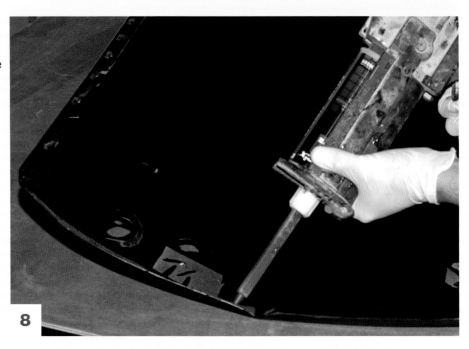

8

REPAIRING RUSTY SHEET METAL

Chris then continues across the bottom of the door skin, all the time, making sure that the adhesive is applied to the surface that will abut the inner panel.

A bead approximately a quarter inch wide is basically centered on the door skin where the surface has been scuffed previously.

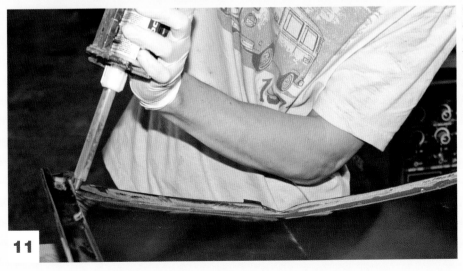

Chris continues to apply the adhesive to the top of the opposite side of the door from where he started. The lip that does not have any adhesive applied actually fits over a portion of the inner door panel.

To avoid contaminating the adhesive with oils from his fingertips, Chris puts on a disposable glove and then uses his fingertips to spread the adhesive.

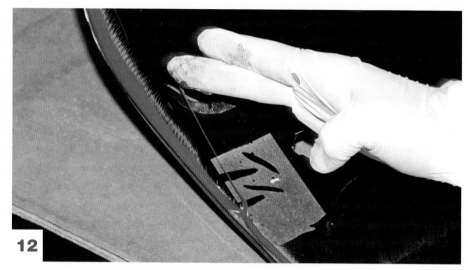

Chris then applies a bit of adhesive on the lip of the inner door panel where the outer door skin will lap over.

With the door skin exterior side down and sitting on a suitable work surface, the inner door panel is slid into place. Make sure that the inner door panel is positioned as far into the door skin as possible and is centered front to back.

The previous steps would not be required if you are not using adhesive to secure the door. With the door skin and the inner door panel adequately positioned in relation to each other, the tedious portion of the repair begins: hammering the edges of the outer door skin over the edge and flat.

15

16

At this point, portions of the door skin are at about 90 degrees to the flange that they will be hammered over, while at other areas, it is at about 180 degrees.

17

Prior to hammering any of the edge over the flange, the door skin should be clamped in place with as many clamps as you have available. You definitely do not want the door skin moving in relation to the inner door panel, and the adhesive is still pliable at this time.

At the minimum, try to clamp the door skin to the inner panel at the four corners. You may have to move some of the clamps while hammering, so having multiple clamps in use will help to prevent movement between the two panels.

18

Besides preventing movement between the two panels, the clamps hold the two pieces together tighter, allowing you to fold the edge over tighter, providing a better seam.

19

Chris will be using a hammer-on dolly approach for folding the door skin edge over the inner door panel flange. In other words, he will be hammering directly onto the dolly, with the two panels in between the two.

20

To minimize marring of the new door skin, Chris has wrapped a shop towel tightly around the dolly. When skinning a door, you do not want the dolly to have any effect on the door skin, but the dolly is required so that the hammer does not simply bounce off the door panels. The shop towel provides some padding to prevent marring, but still provides a solid surface for hammering against.

Chris works his way around the door skin, hammering the edge over slightly with each pass. You should avoid attempting to hammer the door skin edge over completely all at once.

You will notice that Chris's hammer has one end that is straight and another that is curved. This curved end allows the user to strike the surface at a perpendicular angle when in tight situations.

The curved end may also be useful where there is ample room, but when the positioning of the workpiece and the body repairer is at an angle that the straight end of the hammer may require extra effort.

24

With each pass around the door, the skin is folded over a little more.

25

Clearly in this photo, the edge of the door skin is closer to being flat and flush with the inner door panel. To avoid damage to the door skin, several relatively light taps are more productive than fewer heavier hits. One of the secrets to successful bodywork and metalworking in general is practicing finesse.

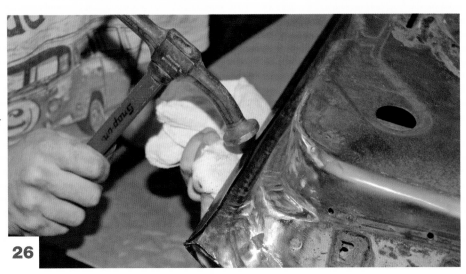

26

Working around the corners of the door will require a little more effort to get the desired result.

Chris keeps the shop towel tightly wrapped around the dolly to help minimize damage to the outside of the door skin.

Special attention will be required where the door changes shape. Take your time, use finesse, and you will learn how to maneuver through these areas like a pro.

Take your time to make sure that the door skin edge is hammered as flat as possible against the inside of the inner door panel. If it is not flat, it will look funny.

30

Coming down the back stretch, this door skin installation is almost finished. Chris has installed lots of door skins, but the process still takes about an hour, so it will most likely take longer for a novice.

31

After all of the hammering has been completed, install as many clamps as possible so that the adhesive cures evenly. The 3M Door Skin Adhesive will require about four hours to cure completely.

32

33

Virtually as good as a new door, but considerably less expensive, this door simply requires normal paint preparation at this point.

34

This is the door for the opposite side of the vehicle. Chris had reskinned it earlier in the day. No dents, no rust, and almost ready for paint.

Chapter 4
Filling, Masking, Priming, & Painting

PLASTIC BODY FILLER

While you can cut out rusty metal and glue on a replacement piece or fiberglass over the hole, chances are very good that you will need to use at least a little bit of body filler to complete the repair. If covering a hole with sheet metal, you will most always need some filler to transition back to the original sheetmetal contour. If you are able to weld in a replacement panel and then metalwork it, you may not need filler, but that is beyond the scope of this book.

If you go the fiberglass route, I'm pretty sure that you are not going to get the fiberglass resin or mat to lay in the exact contour as the original sheet metal. Even if you did, you would need to use at least a skim coat of filler to smooth out the texture of the fiberglass material. Additionally, after coming this far, you might want to repair any small parking lot door dings while you are at it. You may not notice it now, but after you have applied a fresh coat of paint, any surface imperfections will stick out like the proverbial sore thumb.

Mixing Plastic Body Filler

Plastic body filler is sold as two parts: 1) filler and 2) catalyst (a.k.a. hardener) in different size containers. Quart and gallon size cans of filler are very common and come with an appropriate sized tube of hardener. Most body fillers use a hardener that is a distinctly different color than the filler itself, so that it is easy to tell when the two are mixed thoroughly.

Begin by scooping some amount of filler onto a mixing board, clean sheet metal, or piece of flat plastic. Squirt some hardener onto the filler, and then use a plastic spreader to fold the two together. The amount of hardener to use will depend on your shop conditions regarding temperature and humidity. Practice is the best way to determine how much to use, but as a start, add a proportionate amount of hardener to the filler (i.e., a quarter tube of hardener to a quarter of the container of filler). As you begin mixing, you will see streaks

of the hardener color in the filler color. Mix the filler and the hardener by folding the combination onto itself until all of the streaks are gone. If there are streaks of color, you need to continue mixing.

If you apply too little hardener, it will not set up properly. Apply too much hardener and it will set up right there on your mixing board. Yes, it will probably take the entire project before you get the proportionate amounts just right. If you do get it mixed a little too cool, you can speed the curing process slightly by placing a portable heater or heat lamp nearby. If the filler begins to "kick" before you have it spread out, you might as well scrape it off the mixing board and throw it away.

Applying Plastic Body Filler

Most all body fillers are applied using the same methods, but you should read the directions for the particular product that you are using to be sure. Typically, the surface to be filled is sanded down to bare metal before filler is applied. Note that some fillers suggest that the surface be stripped of any paint and a coat of epoxy primer applied prior to any filler being applied. Most auto body paint and supply stores will be able to provide printed information telling you specifically which products are and are not compatible.

After the filler is thoroughly mixed, use a clean plastic spreader to spread the filler over the area to be filled. You don't have to be perfect at this point, but the better you spread the filler, the less sanding will be necessary. However, you should keep in mind that you will be sanding this filler material, so it does need to be a little high. Of course, you can add more later if necessary.

If it starts too low, you are assured of needing to apply multiple coats. Do all you can to avoid having any plastic filler that is thicker than ⅛-inch thick when it is finished. Since you are not building a show car here, you can make it a little thicker if

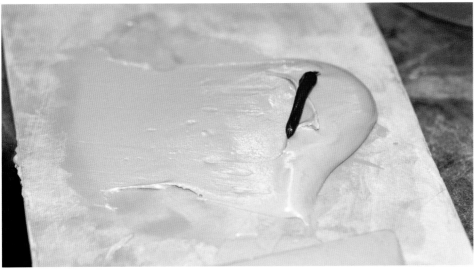

Squeeze the recommended amount of hardener onto the filler. Estimating the correct amount of hardener will become easier throughout the project, but it will be a trial and error process when you first begin. Typically, a proportionate amount of hardener is used with the filler (in other words, ¼ of the hardener should be used with ¼ of the filler).

Using a plastic spreader, fold the filler over onto itself and into the ribbon of hardener.

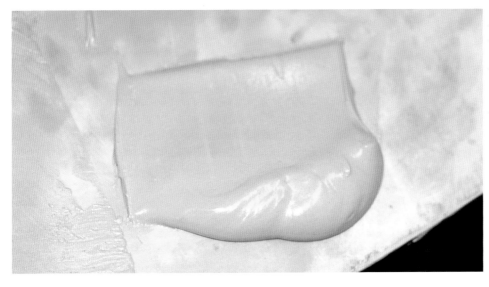

Using the spreader, scrape the filler and hardener away from the mixing board, and fold it onto itself again. Continue using this folding over motion from each side until the filler and hardener are thoroughly mixed. The two are thoroughly mixed when the two are one consistent color with no streaks.

Using a clean spreader (use the largest size that will fit into the area being filled), spread the filler onto the surface to be filled. No filler should be more than ⅛-inch thick. Then go back with a plastic spreader and smooth out the filler, eliminating ridges in the process. As you mix and use additional filler, you will no doubt learn to use more or less hardener based on how long it takes you to get it spread out and how long you have to wait before you can begin sanding it.

necessary, but you should build the total amount up in thinner layers rather than all in one to ensure that the inner layers thoroughly cure. Filler, like most auto body repair products, cures as its various chemical components react and escape from the material that is left. If the filler is applied too thick, it will quite often cure on the outside before all of the chemical reaction has taken place on the inside, trapping uncured material inside of the repair. When this happens, the repair won't last like it should and will ultimately show up in the finished paint job.

You should also verify that you do not create any voids in the filler. As you spread the filler, you should use a smooth, even pressure stroke to move it where you want it. If you get going fast and furious, the filler could bridge over itself, creating a void beneath. If this does happen, you would most likely sand through it

and simply need to apply a second or third coat of filler in that area. However, a worse case would be that you might not sand into it, leaving a thin coat of filler that could easily be poked through later.

Sanding Body Filler

Sanding blocks are used to exert even pressure on the sandpaper, so that you can minimize waves in the panel being sanded. Since automotive body panels come in a variety of contours, sanding blocks have different requirements. If the panel has lots of curves or otherwise round surfaces, the sanding block needs to be flexible, but firm, to maintain even contact between the body surface and the sandpaper. To be flexible, sanding blocks are made out of rubber or various types of foam. To fine tune the flexibility of some sanding blocks, some are designed with removable rods that slide into or out of the back, stiffening the block with more rods, making it more flexible with less.

On the other hand, if you are sanding a flat surface (such as a hood or roof panel), a more rigid sanding board would be appropriate. The more rigid and longer the board, the more effective it will be in eliminating waves in the panel. A longer board will span across multiple ripples, working to knock down the high spots, rather than simply skimming across the surface as a short sanding block would do.

Some older types of filler require initial smoothing with a cheese grater–type file, while most newer products can be smoothed initially with 80-grit sandpaper. However, you should check with the person behind the counter where you purchase your products to determine the best method. If you are using any type of filler that requires using a cheese grater, you will soon realize that the initial smoothing should be done slightly before the filler cures completely. You can watch the edges of the filler to get a feel for whether it has cured enough or not. If the filler starts breaking away at the edges or if the sandpaper starts loading up, the filler has not cured sufficiently. It is difficult to describe the correct time, but with a little bit of practice, you can quickly get a feel for it.

You want to knock off the high spots before the filler gets rock hard, but not too soon or you will easily

APPLYING BODY FILLER OVER A FIBERGLASS REPAIR

After adding some fiberglass mat and cloth to restore some strength and rigidity to a rust-damaged fender, it is time to make it more cosmetically appealing. Depending on your particular vehicle and how much time and effort you want to spend, you can do this quickly, or you can take some time to do it to the best of your ability.

The bulk of this work will be spreading plastic body filler, sanding it smooth, filling low spots, and repeating until it has the desired contour and finish. Contrary to common belief, bodywork is what makes or breaks the finished paint project. If the preparation work is well done, the paint will look good. If the preparation is done poorly, no amount of paint will make it look any better.

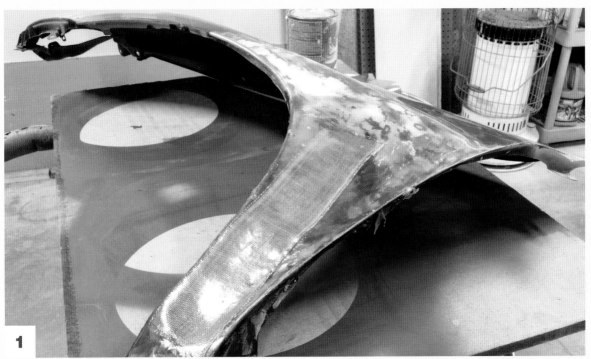

1

Much of this fender was caved in due to collision damage that, left unattended, developed into rust. Since the damage was too deep to fill with plastic body filler and seemingly too thin to hammer back out, it was filled in with some strips of fiberglass mat and fiberglass cloth. It just needs some plastic body filler, primer, and paint to make it look presentable.

2

Using a mixing board, cardboard, or, in this case, a scrap piece of sheet metal as a palette, scoop some plastic body filler onto it and add some hardener. Then use a body filler spreader, putty knife, or other flat object to fold the two components together until there are no streaks of color. The filler should be uniform in color or it is not mixed thoroughly. If working on a larger area, it may be beneficial to not add quite as much hardener, providing you with more working time.

3 Use a clean spreader to apply the body filler. Depending on the filler that you use, it will have a consistency similar to peanut butter, allowing it to be spread rather easily. The goal will be to spread it as evenly as possible, although it will need to be feathered back into the original surface around the edges. When working in a small area, or if the shop temperature is low, it may be beneficial to add a bit more hardener.

4 After just a few minutes, this batch of filler was ready to sand smooth. With a new sheet of 80-grit sandpaper in the sanding block, commence to sanding in as many different directions as possible. This is a narrow stretch of fender, so the direction is somewhat limited. If the sandpaper starts clogging up, wait another minute or two before proceeding.

After just a bit of sanding, the filler is not as shiny and the high spots are quickly getting smoother. The first coat of body filler will usually be the worst as you are typically not applying it to a smooth surface. Once the initial layer of filler is sanded smooth, successive layers will naturally go on smoother.

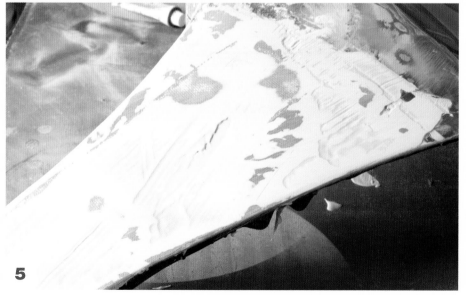

5

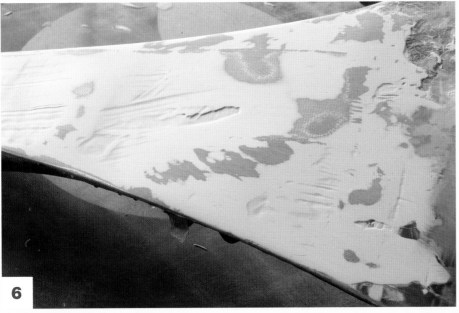

The gray spots are bare metal, indicating that no more filler should be put in that specific area. More filler will need to be added to the left (bottom of fender) and to the right (top and rear of fender). While some would apply more filler to these areas at this time, my personal preference is to sand the entire area smooth, regardless of how much more filler is required. Again, filler is easier to apply to a smooth surface.

After a bit more sanding, a second layer of filler was added. Most of this will be sanded off, but there are areas to be filled.

Although it isn't necessary, you can speed up the process of removing obvious ridges of unwanted body filler with a cheese grater file. If you catch the filler at the right time, it will come off just like cheese, in thin strips. With the improvements in plastic body filler over the years, most of it can be smoothed with sandpaper quite easily ... unless you let it cure way too long.

Using the cheese grater file, the high spots can be knocked down quite easily. However, this is merely removing material, not smoothing it at all. It will still be necessary to sand it down to the correct contour and surface finish.

9

Although more sanding is still required, progress has been made. Remember from a previous photo that the area roughly between the pink body filler can and the bare metal spots at the back of the fender are high, so by the time this filler is sanded smooth, it might be close.

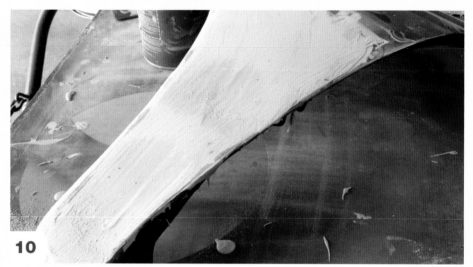

10

Some low spots are still present, but it is getting late in the day. Some more filler, as well as more sanding, will be required. To ward off any formation of rust on the bare metal spots until the next work session, some primer will be applied.

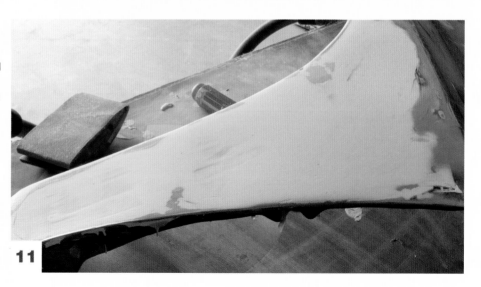

11

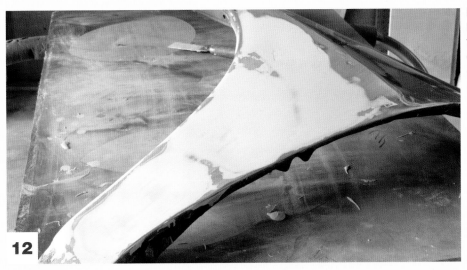

Prior to applying any primer, however, the surface needs to be cleaned again. Spray or wipe on some wax and grease remover with a clean paper towel.

12

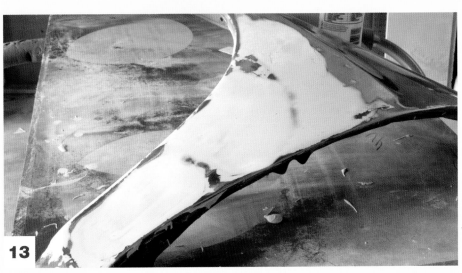

Then wipe it off with another clean paper towel. While the fender may look smooth in the photo, imperfections are still prevalent up close. They will not look any better when primer is applied.

13

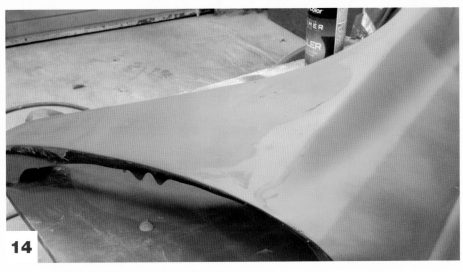

See, I told you … Besides a bit more filler in the low spots, what is there needs to be feathered in to the original surface more. Also, the approximately 1-inch-wide body line around the wheelwell needs to be reestablished from about the middle of the wheelwell back.

14

gouge out more material than you really want to. As you begin working the filler, sand the entire filled area first with 80- or 100-grit sandpaper, then switch to 200- or 240-grit to blend the filler into the surrounding area. When you are finished sanding with 240-grit, you will have a good idea if more filler is necessary or not prior to applying primer.

When finished sanding, blow all of the sanding dust away with an air nozzle. If low spots are remaining, lightly rough up the area to be filled with the previous grit of sandpaper, and then mix an appropriate amount of body filler and apply it as before. Work the second and successive layers of filler (if required) just as the first until any and all low areas are filled.

You can use 80- or 100-grit sandpaper to smooth the filler, but a cheese grater file works better and faster. Some of these files are flat, while others have a rounded shape. The rounded ones seem to be more durable and less likely to break. The makers of these files make handles for them, but most auto body repairers who I know seem to use them without a handle. The filler requires a few minutes to set, but you should begin sanding or filing before it sets completely. When the consistency is correct, the filler will come off in strips just as cheese would come through a cheese grater.

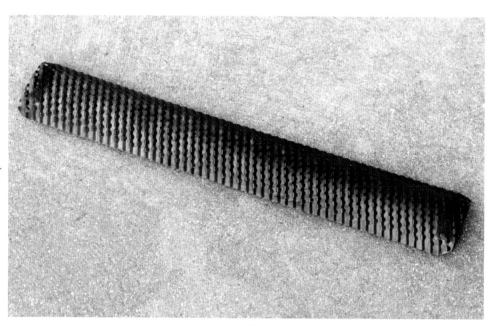

When using a cheese grater file, you are just concerned with taking off the rough edges or high spots off the filler. It is much smoother than originally, but still lots of block sanding must be done. With the high spots knocked off, break out your favorite long board sanding block and some 80- to 100-grit sandpaper. Sand all of the fiberglass-reinforced filler until it is smooth. By using the longest sanding block that will fit in the area, you will be working toward getting the area flat (not wavy), as well as smooth. More filler will cover this layer, but the sooner you get the filler flat the better.

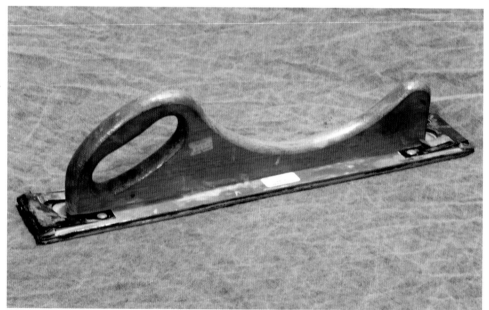

Occasionally, you should use an air hose with medium pressure to blow away any buildup of body filler from the sandpaper. Be sure to direct the air pressure away from you or anyone else in the area.

Although most sanding blocks and boards have a flat surface, some sanding boards are designed with a concave surface, allowing you to smooth the inside of a curved surface. These are available in a variety of radii.

Whenever you purchase sanding boards or blocks, you should take special notice of how the sandpaper is held in place, as not all sandpaper is compatible with all sanding boards and blocks. Small inexpensive rubber sanding blocks often have a horizontal slit in both ends, with the top flap having two or three sharp tacks that grab the sandpaper. The sandpaper is wrapped around the bottom of the block and the ends of the paper placed between the upper and lower flap and held in place by the tacks. Some sanding boards have spring clips that hold the paper in place. Still others use adhesive-backed sandpaper.

Sanding Body Filler Repairs

Regardless of the type of surface you plan to paint over, whether it is body filler or an existing paint finish, some sanding will be required. This phase of any automobile paint operation is just as critical as any other. Remember that every blemish or surface flaw will be magnified by paint coats.

Top layers of body filler are initially sanded with 80- to 150-grit paper to smooth and flatten rough spots and to get the surface close to an even texture. Then, 240-grit paper is used to make that finish even smoother and flatter. Sanding must be done with a sanding board or block. After every two minutes or so, feel the surface with your open hand to judge your progress. Any irregularity you feel, you'll see with paint on it, so keep sanding and checking the area until it's smooth and flat and perfectly blended with surrounding surfaces.

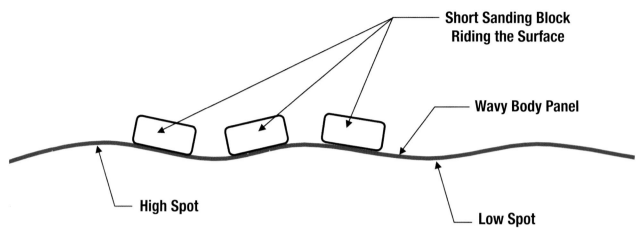

Short Sanding Block Riding the Surface

Wavy Body Panel

High Spot

Low Spot

Besides smoothing the surface, part of the reason for sanding is to make wavy panels flat. A short sanding board or block can be used to get any panel smooth, but it will not necessarily get the panel flat. A short sanding board or block will merely ride over the ridges, rather than knocking them down.

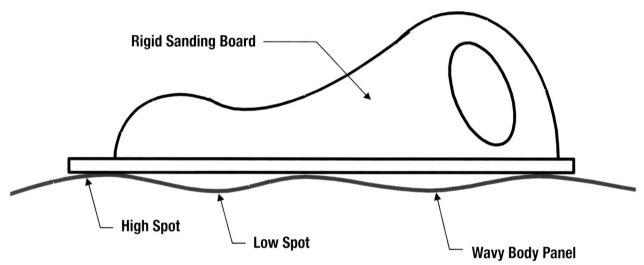

Rigid Sanding Board

High Spot

Low Spot

Wavy Body Panel

On the other hand, by using the longest sanding board possible for the area being sanded, the sandpaper will span across the ridges in a body panel and will quickly knock down the ridges, making the panel flat as well as smooth.

Operate sanding boards and blocks in all directions. Do not simply maneuver them in a back and forth direction from the front to the back. Move them up and down and crossways diagonally, rotating the board or block as necessary for ease of operation. This multidirectional sanding technique will guarantee that all areas are sanded smooth without grooves or perceivable patterns.

Once you've made the area flat with 240-grit paper so there are no remaining high spots, wrinkles, grooves, or ridges, use 320-grit paper to remove lingering sanding scratches and other shallow imperfections. Up to this point, you've been shaping the body filler until it is flat and it blends with panel areas adjacent to it. Now, with finer sandpaper grits, you will focus on texture smoothness.

When satisfied that your filler repair has been sanded to perfection, use 320-grit paper to gradually develop a well-defined visual perimeter around the entire repair area. This "ring" around the repair should expose a band of bare metal about 1 inch wide and then successive bands of equally wide exposed rings of primer, sealer, primer, and paint. Because undercoat and paint products consist of different colored materials, you will be able to see your progress clearly. The object is to develop sort of a layered valley of smooth walls between the top surface of the body filler area and the top surface of existing good paint. This allows fresh applications of undercoat material to fill to the same thickness as those same materials existing on the rest of the car's surface.

The next step would be to apply a coat of primer to the repaired surfaces. Primer will make any imperfections somewhat more noticeable, and therefore, indicate trouble spots sooner than later. After allowing the proper drying time for the primer, break out your favorite sanding boards or blocks and a supply of 400-grit sandpaper. Using a crisscross pattern, sand the entire area that has been repaired with 400-grit sandpaper. When you have sanded the entire area, look for any places that might need just a bit more filler. Apply the filler as described previously, then re-sand this newly filled area with 320-grit and then 400-grit sandpaper.

This approach will allow you to apply final color coats in thickness equal to the rest of the paint finish for the best possible blend, color tint, and texture identical to surrounding paint finishes. This process is referred to as *feathering in*. Subsequent coats of primer material will also be sanded to a point where the only depth difference between an existing painted surface and a repair area will be the actual thickness of the existing paint.

Sanding Existing Paint Surfaces

Applying new paint over old paint without properly scuffing up the old surface is a mistake, especially when dealing with factory paint jobs that were baked on at 450 degrees Fahrenheit. Situations like this commonly result in new paint flaking off because it does not have an absorbent base to adhere to. The super hard baked-on paint jobs do not always allow new paint to penetrate their surfaces.

Fine-grade Scotch-Brite pads work great for scuffing baked-on paint finishes. The comparatively rough finish left behind makes a great base for coats of

sealer. You could also use 500- to 600-grit sandpaper to scuff shiny paint finishes. The overall purpose is to dull all shiny surfaces so that new layers of material have something to grab onto. There is no need to scuff or sand in one direction only. You can, and should, sand in all directions to be sure all surface areas have been roughened up satisfactorily.

Cleaning Sanded Surfaces before Undercoat Applications

Once you've sanded or scuffed the surface as required, you'll need to clean it thoroughly to remove all surface contaminants. Painters normally use air pressure to blow off layers of sanding dust from body surfaces, as well as between trunk edge gaps, door edges, and doorjambs.

Next, they use wax and grease remover products to thoroughly wipe down and clean body surfaces. Each paint manufacturer has its own brand of wax and grease remover that constitutes part of an overall paint system. You should use only those wax and grease remover products deemed part of the paint system you will be using.

Dampen a clean cloth (heavy-duty paper shop towels work great) with wax and grease remover and use it to thoroughly wipe off all body surfaces in the area to be painted. The mild solvents in wax and grease removers loosen and dislodge particles of silicone dressings, oil, wax, polish, and other materials embedded in or otherwise lightly adhered to surfaces. To assist the cleaning ability of wax and grease removers, follow the damp cleaning cloth with a clean, dry cloth in your other hand. The dry one picks up lingering residue and moisture to leave behind a clean, dry surface. Use a new towel on every panel, wipe wet, and dry thoroughly.

To ensure super clean and dry surfaces, go over finishes with an aerosol glass cleaner after a wax and grease remover. The ammonia in such glass cleaners helps to disperse and evaporate moisture, as well as to pick up missed spots of wax or dirt residue. Instead of wetting a cloth with glass cleaner, spray the material on surfaces and wipe it off with a clean, dry cloth.

When you are ready to learn more about automotive bodywork, please check out my book, *The Complete Guide to Auto Body Repair, 2nd Edition.*

RESTORING A FADING BODY LINE

Sometimes, especially when repair is part of the equation (compared to simple paint preparation), some of the vehicle's body lines may be damaged or missing. This can make it very difficult to accurately restore a damaged panel, especially if you do not have a similar panel to compare it with.

On the front fender shown in various stages of repair, there is an approximate 1-inch-wide flat lip around the wheelwell. With the fender mounted on the vehicle, the surface of this lip would be close to vertical in orientation. Past the one-inch-wide lip, the sheet metal turns inward at a slight angle. There should not be any perceivable step in the surface if you were to run your fingernail across it, merely a difference in direction.

This lip is still intact from about the middle of the wheelwell forward, but disappears moving rearward. With the damage that the fender sustained and the anxiousness to get the fender repaired, it can be easy to concentrate on sanding to the point that the lip never gets restored. Not all vehicles have this type of lip around the wheelwells, so it may not even be an issue. However, if part of it is there, while the other part is missing, it will look goofy.

To restore this missing body line, the first step will be to outline the area to be filled with ⅛-inch 3M Fine Line tape, which will be followed by some wider masking tape. Using that as a border, a bit of body filler will be added all along the wheelwell where the lip is missing. That lip area will then be sanded flat from the wheel opening side, the tape removed, and then the rest of the fender blended in.

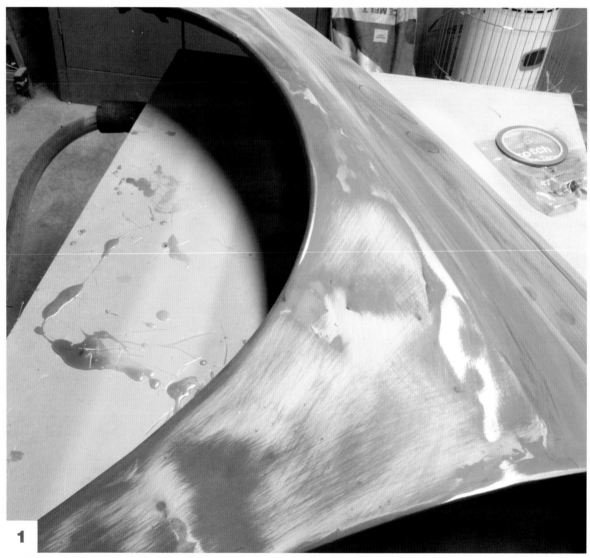

1

As you are sanding, you may find that it is easy to sand out a body line, whether you want to or not. This is one of the things that separates the hacks from the advanced amateurs.

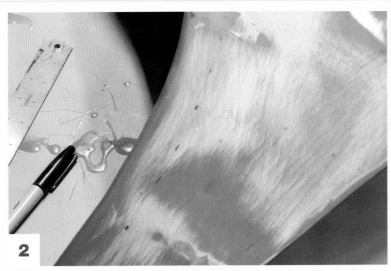

Here we can see that the surface is relatively flat (not perfect), but not sporting the original body line around the wheelwell opening either. A casual observer might not even notice, but if you pointed it out to them, it would quickly stick out like the proverbially sore thumb. A permanent marker was used to lay out a series of dots the same distance in from the wheelwell.

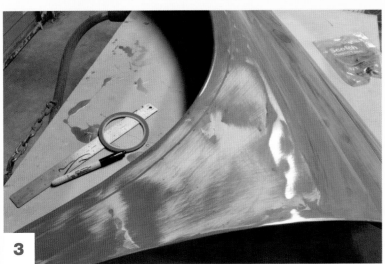

Fine line tape was used to connect the dots, providing a visual guide as to where the lip should be. While the damaged lip might not be quite as crisp as the original stamping, the measurement might leave a bit of a zig-zag line. You should try to make the tape follow as smooth of a curve as possible, even if it varies from your measurements somewhat. The ⅛-inch tape curves much more easily than regular masking tape.

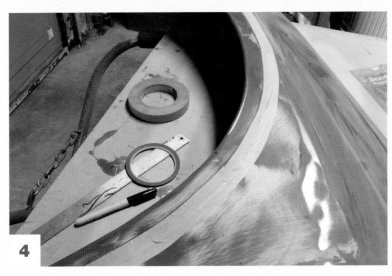

After laying down the fine-line tape, a strip of ½-inch-wide masking tape is placed so that it overlaps the fine line tape somewhat. The fine line is still going to be exposed on the edge that will be filled. A second layer of ½-inch masking tape is also applied.

Before mixing any filler, check for any other places that might need some filler. The circled areas are just slightly lower than the surrounding sheet metal as the sanding has revealed, but since we are mixing some filler anyway, there is no reason not to fill them.

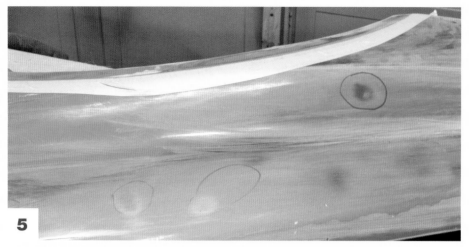

5

Body filler was first applied along the wheelwell lip, and then other low spots filled.

6

After allowing the body filler to set up a bit, sand the wheelwell-lip rim so that it takes on the desired contour. Depending on the vehicle, this may be a narrow flat band, a wider flat band, or sometimes a bit of a reverse curve. Whatever it is, the intent is to reestablish the original contour as closely as possible.

7

After the wheelwell lip has been properly shaped, remove the masking tape and the fine line tape. Then sand any other filler that has recently been applied, blending all of the panel and its body lines together so that there are no perceivable steps in the surface. Remember that the edge of this lip should really just be a line indicating the intersection of two planes of the fender's surface.

After shaping the body filler with 80-grit sandpaper, verify that there are no low spots. If there are, fill them now, then sand it all smooth with 120-grit sandpaper over the entire surface. Then repeat the sanding process with 220- and 320-grit sandpaper.

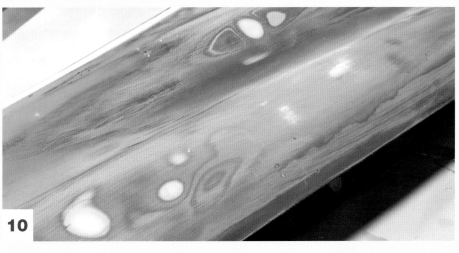

Notice how none of these spots of body filler have had lines around them. Instead, they have somewhat fuzzy edges indicating that the low spots have been feathered in. In a professional paint job, the area between primer, filler, and previous top coat would be much broader, giving an overall smoother finish.

After yet another coat of primer, I still was not satisfied with the contour of the repaired area. With this area being low, I could not achieve the wheelwell lip contour that I desired. If you are not happy with the work at this point, spraying paint on it is not going to make it any better.

11

By filling in the remaining low spots in the narrow area, it was easier to reproduce the wheelwell edge, making it look more like it did from the factory.

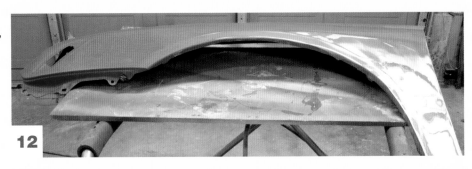

12

The wheelwell edge seems to be better delineated now. By now, you should be pretty intimate with the panel in question. Low spots should be filled and sand scratches sanded out with finer sandpaper. The panel should be ready for the last coats of primer at this point.

13

MASKING BEFORE SPRAYING

It's difficult to paint around windows without getting overspray on belt moldings and trim pieces surrounding the glass. The surest way to avoid overspray is to remove the glass and trim. Barring that, use plastic tape to outline outer molding edges next to those panels to be painted. Attach strips of wider masking tape and paper anywhere along the plastic tape's width; just verify that there are no gaps between the inside edge of the fine line tape and the wider tape.

You only need one strip of masking paper to cover windows, as long as it is wide enough to reach from top to bottom. Fold paper as necessary to make it fit neatly along the sides. Use strips of tape to hold the paper in a secure position and to cover any resulting gaps. A single sheet of automotive masking paper will prevent paint from bleeding through to underlying surfaces. If you have chosen to use a masking material other than recommended automotive masking paper, you might have to apply two or three layers.

Masking paper does not always come in widths that fit window shapes exactly. Most of the time, especially with side windows, you end up with a tight fit along edges and bulges in the middle. To avoid bulges, fold excess masking paper so that it lies flat. Not only does this make for a tidy masking job, it prevents bulky paper from being blown around by air pressure from a paint spray can. All you have to do is lay one hand down on an edge of the paper and slide it toward the middle. With your other hand, grasp the bulging paper and fold it over. Use strips of tape to hold it in a neat fold.

Whenever masking, always remember that paint will cover everything that it touches. Very small slits between tape and masking paper will allow paint to reach the surface below. Lightly secured paper edges will blow open during spray paint operations and allow mists of overspray to infiltrate underlying spaces. Therefore, always run lines of tape along the length of paper edges to seal off underlying areas completely. This is especially

Since this is fender is being repaired off the vehicle, there is nothing to mask, except for anything else in the shop that you don't want to get overspray on. This mounting bracket can be removed far more easily than masking it properly. Having even the slightest bit of primer on it would look amateurish, so it is best removed.

Remove two Phillips head screws with a screwdriver and no masking is required.

Removing this easily removed and replaced bracket also allows for complete scuffing, priming, and painting of the fender. The marker light had been removed some time ago; otherwise, it would be removed at this time also.

important when the edge of one piece of masking paper is overlapped with another.

Windshields and rear windows are generally quite big. You might have to use two or three strips of paper horizontally placed in order to cover all glass. If you leave trim pieces in place, consider applying fine line tape around their outer edges before working with wider tape and paper. You must remove rear window louvers or side window air deflectors when painting body areas next to them.

Rubber moldings that lap against body panels—rubber windshield moldings along roofs, A-pillars, and cowlings—present difficult masking challenges. To make the job of placing tape directly over the molding's

MASKING A WINDSHIELD

1 This type of windshield installation is common on many contemporary vehicles. There is no rubber molding to contend with, yet there is more than glass to be covered. Begin by running a piece of masking tape along the edge of the body, covering what should not be painted, but leaving the part to be painted exposed.

Using 1-inch-wide masking tape (narrower tape is a little easier to work with in areas that are not straight), cover both A-pillars and across the top of the windshield.

2

Windshield wipers fit into the recess at the bottom of the windshield and behind the hood, but you do not want them being painted. You can remove the wiper arms, but you still need to mask the recess that they fit into, so in this case the wipers were left on.

3

Since the masking material is transitioning roughly 90 degrees on each side of the windshield, secure a second strip of tape onto the first piece along the A-pillars over to the glass. Using the widest tape you have available will make this easier. Also attach a piece of masking tape to the back edge of the hood (if it is not going to be painted) or to the underneath side of the hood or the plastic trim. This first piece of trim along the back edge of the hood will provide an attachment point for the masking paper that is to follow.

4

Cut a piece of masking paper long enough to span the lower portion of the windshield. Since two pieces will be required to fully cover this windshield, position the first so that it reaches the tape at the back of the underneath side of the hood. Then tape it in place across the windshield so that it doesn't move.

Now trim the edges and bottom as required so that they do not cover any portions of the body that will be painted. Press the paper as flat as possible, refold if required, then tape securely in place.

Following the same general principles, apply a second piece of masking paper to cover the top of the windshield. Tape the paper to the tape already along the top of the windshield, then fold the excess on one end to fit the windshield, and tape it down.

Press the paper flat against the windshield and trim or fold the excess to fit. Then tape down the end of the paper, along with any flaps that may remain.

If there are any gaps between the two pieces of paper, cover them with another piece of paper or the required number of strips of masking tape if the gap is a relatively small area.

Press down all of the tape, verify that all seams are covered, and that there are no gaps. If you follow this procedure, it makes removing the masking simple. Begin at the first corner where the first piece of tape was applied and pull firmly, but gently, and the masking can often be removed in just two or three pieces.

outer edge easier, consider putting a length of thick, non-scratching cord under it.

The Eastwood Company carries a weatherstrip masking tool designed to insert long strips of plastic cord under the edges of molding and weather stripping. It raises these edges up off body surfaces to allow complete masking under them. This way, paint can reach under molding edges, instead of just up to them, ensuring good paint coverage and eliminating distinctive paint edge lines next to moldings.

If you've removed the windows and moldings, you'll have to mask off the interior compartment. Generally, exterior paint colors are applied to the middle of window openings. You could lay down strips of wide tape inside these openings and fold it over toward the inside for side windows. Along the spot-welded metal edge of windshield openings, simply apply perimeter tape and paper from the interior compartment side.

Emblems and Badges

As with trim pieces, it is best to remove emblems and badges before painting. They are secured by clips, pins, screws, adhesives, or double-backed tape. Be extra cautious while attempting to take these items off any vehicle. Too much prying pressure will cause them to break. Unless you can see that their protruding support pins are secured from inside a trunk space, inner fender area, or other locations, you will have to carefully pry open an edge to determine just how they are mounted.

If you are not sure how to remove those items, consult a dealership service representative, auto body paint and supply jobber, or professional auto body painter. Should an emblem or badge break during dismantling, don't despair. Even professional auto body technicians break these plastic items occasionally.

To ensure that emblem and badge edges are completely covered, mask carefully around their edges, allowing no part of the tape to extend onto the painted surface. Tape over the edges first, before masking their faces. Again, 3M Fine Line tape may be the best material for this meticulous task. After attaching the tape's end to a corner of an emblem, maneuver the roll with one hand while carefully placing and securing the tape with your other hand. Practice is essential, so do not expect to accomplish this kind of unique masking on the first try.

Some painters make this job easier by covering emblems with wide strips of tape first. They then use a sharp razor blade to cut tape along the emblem edges at the exact point where they meet the painted body. You must use a very delicate touch to avoid cutting into paint or missing the mark and leaving an open gap along the part being masked off. If you should decide to try this technique, opt for very light passes with the razor blade, even if it takes two or three attempts to cut completely through the tape. This will allow you to avoid a deep scratch in the paint, should your hand slip.

Door Locks and Handles

Because door locks and handles are secured right next to painted door panels, the same kind of meticulous masking is required for them as for emblems and badges. The best approach is to remove them, unless you're doing only clear coats or light paint feathering or melting in up to their edge.

Door handles are best masked using tape for the entire process. Paper, even in 4-inch widths, is just too cumbersome to work with. Use ¾-inch tape to mask the perimeter edges and then 2-inch tape to completely cover the unit. If your car presents a rather unique handle, employ whatever means necessary to cover it. Use your imagination. Tape can be applied initially from the back to offer sticky edges that can extend out past upper and lower edges and can be folded over to cover the front. Remember, the most critical part of masking is along the edge, where items meet painted panels. Wide strips of tape can easily cover faces and other easy-to-reach parts.

Key locks are easiest to mask by simply covering them with a wide strip of 1- or 2-inch-wide tape, and then cutting the excess from around the lock's circumference with a sharp razor blade. Before cutting, though, use a fingernail to force tape down along the circumference to be sure coverage is complete and that the tape is securely in place.

Other Exterior Features

Remove the radio antenna if it protrudes from a panel you're going to paint. They often unscrew from their base, leaving a large gap between the car body and remaining antenna unit. If you decide to leave the antenna in place, don't wrap it barber pole style from top to bottom. Instead, sandwich it between two vertical strips of tape—they'll cover it just as well and be much easier to remove.

MASKING A REAR WINDOW

The author's 2008 Chevrolet Silverado rear window is among the type of vehicle where the rear window is glued into place, without any weather stripping around it. This makes it fairly easy to mask should the need arise. It doesn't really matter where you start, but the edge of the glass should be masked off first. I used 1½-inch-wide masking tape, wrapping it inward over the edge and then flat onto the surface of the glass.

A professional might be able to tape the edge of the entire window with one piece of tape, but I did the sides and top with two pieces. It doesn't really matter how many pieces of tape or masking paper you end up using, as long as you get the job done.

In similar fashion, tape across the bottom of the glass. Verify that you have not covered any of the area on the cab that will be painted. Rub your hand or fingers over the edges to ensure that the tape is stuck to the glass. This perimeter of tape will make it easier to secure masking paper.

Having a masking paper rack that automatically adds a strip of masking tape along one edge as it dispenses makes masking much easier, but I do not have one. So, I cut a piece of 18-inch-wide masking paper long enough to cover a little more than half of the back glass. With a couple of short pieces of tape, I attached it to the tape along the top edge of the glass so that it will be able to cover the end of the glass.

As shown in this photo, the 18-inch-wide masking paper is more than adequate for this rear glass. Wider paper is available if necessary, or you can use multiple rows across the glass. Fold the end of the paper as required to cover the shape of the glass on the end.

Tape the end of the paper down and then fold the excess paper at the bottom upward. Rub your hand over the folds to crease the paper (making them easier to tape over). It is unnecessary to keep the paper precisely flat, but you should attempt to avoid bubbles or bulges that could easily be ripped open and ruin your masking job.

Cover the remaining glass by using the same procedure.

Now that paper covers all of the glass, apply another row of tape over all of the paper edges and the seam in the middle.

Even the excess masking paper that is folded over at each end must be taped down. The primary goal is to prevent primer, paint, or clear coat from landing on the glass. However, a secondary goal is to remove any chance of forming pockets that collect dirt or debris that could fall onto your fresh paint when the masking is removed.

Taillight and side light units are usually quite easy to remove by loosening four to eight nuts on their housings' back sides. Should you decide to mask them instead, use overlapping strips of 2-inch tape. Be sure to overlap each strip by at least ¾ inch to prevent paint seepage into seams.

Bumper designs range from out-in-the-open pickup truck step bumpers to closely fitted wraparound urethane-faced models. You must remove them to paint surrounding areas.

If you're not painting the surrounding surfaces, mask bumpers with regular masking tape and paper. If possible, insert a strip of tape between body panels and those parts touching them. Fold it over and then attach paper and tape to it. Fold paper over the top and face of bumpers. If the part requires more paper coverage, start from the bottom and fold over the top to overlap with the preceding piece.

If you intend to touch up only a quarter panel, and only the side-mounted bumper guard is in your way, consider dismantling the guard alone. These are attached to bumpers with nuts and bolts. They are flexible, for the most part, and bending them out of the way should give you enough room for dismantling them. If not, maybe that piece can be pulled away and secured with tape so you can paint below and around it.

License plates and their holders are very easy to remove. If set into a housing, remove the entire unit if possible. Masking these units requires the same techniques described above for complete coverage with good overlap to prevent seepage.

Vinyl graphics, stripes, and decals are not easy to remove or store. The procedure for removing them typically destroys them. If you can't replace them, you'll have to mask them off when painting near or around them. Use thin fine line tape for a precision line with low paint buildup along its edge.

Meticulously mask the outer perimeter edges of vinyl graphics, stripes, and decals first. Be sure to place fine line tape directly on top of the item being masked and perfectly in line with its edge. When touching up or spot painting a body section below such items, you need only to mask the bottom edge. The rest can be covered with paper and regular masking tape.

Grille

If you don't need to remove the grille, mask it with wide strips of paper. Attached to the top of the unit first, paper will hang down to cover most of the assembly. Use ¾-inch tape to secure paper on the sides. There is no need to mask individual contours or sections. If you have 12-inch paper and need to cover a grille that is 20 inches top to bottom, simply tape the paper edge to edge and attach the extra-wide sheet as a unit.

Intricate painting next to the grille requires that the grille be removed. No amount of masking will allow you adequate spray gun maneuverability. However, if all you need to do is paint the front parts of your ground effects system, then start masking from the bottom of the grille and work upward. When masking, always keep in mind the painting requirements for the job and the sequence you plan to follow.

The most important part of a masked area is the section adjacent or perpendicular to the area to be painted. It is the edges of those items that will expose overspray and the paint buildup edge so predominant with repaint efforts. Be sure the first piece of masking tape placed along the edges of those items is secure, adequate, and accurate.

Unaffected Body Areas

Before you begin your project, plan to mask every part of your car that will not receive primer or paint.

This doesn't mean you have to mask individual trim items on the driver's side when all you will be painting is the passenger side quarter panel. You will need to mask the driver's side of the car, however, or overspray will settle on it.

At most body shops, painters use large sheets of plastic to cover everything beyond the immediate painting area. They hold the plastic in place with tape so that air pressure doesn't blow it around and knock it loose or stir up particles.

SPRAYING WITH SPRAY CANS

By now you are getting ready to spray your final coats of primer and then color, so you want to make it as good as possible. You have probably already used spray paint somewhere along the line, but you should remember

Filler primer is somewhat thicker than ordinary primer and is intended to fill deep scratches. When compared to paint, all primers are going to be easy to sand, as they are not designed to be a final coat of protection. In all actuality, any primer that is not an epoxy is actually very porous. The bad part of this is that it offers no corrosion protection.

Compared to filler primer, sandable primer is somewhat thinner, filling only the lightest scratches. It should be applied only after all of the bodywork is completed and sanded smooth, to promote adhesion and provide a uniform surface.

a few things when using a spray can, spray bomb, or rattle can:

- Shake the spray can vigorously for about three minutes before using.
- Test the spray tip by spraying on a scrap piece of paper or cardboard, before spraying on the workpiece.

- Spray around the edges of the workpiece first, then spray the rest of the area to be painted. This prevents excessive overspray around the edges.
- Spray a light mist coat first, then cover the entire surface on the second layer. Then apply a complete second coat. Allow to dry thoroughly before moving the workpiece.

Prior to applying that final coat of primer, check all of the bodywork one last time. Are there any low spots or scratches that won't be coverded by a coat or two of sandable primer? If there are any imperfections that you cannot live with, address them now.

- Avoid moving the spray gun or spray can in an arc or pendulum-like manner. Keep the spray tip parallel with the surface at an even distance throughout the pass. Move your body as necessary to maintain a consistent distance between the paint dispenser and the workpiece.
- Overlap each pass by about 50 percent.
- To clear the spray cap when finished spraying, invert the spray can and depress the cap.

Primer

Primer has two main objectives: to cover any body filler beneath it, and to prepare the surface that will ultimately receive paint. While a trained eye can spot bodywork imperfections quite easily, a novice simply doesn't see the imperfections in a coat of body filler. While the filler may be sanded perfectly smooth, it may not be "flat." Like the distorted mirrors at amusement parks that are wavy, wavy bodywork will distort the reflections in the finished paint job. Even the most expensive paint, sprayed from the finest equipment will not hide wavy body panels.

A coat of primer every now and then while doing the bodywork (sanding) will provide you with a better

Available in a wide variety of colors, spray paint offers and quick and easy method of applying color to bodywork repairs.

SPRAYING RATTLE CAN PRIMER

Okay, good or bad, we're getting down to almost being finished. Don't expect your repaired fender, door, or whatever to be perfect. If it is, that is great, but you must remember that the main goal was to make the repair good enough that the vehicle looked better than it did before you fixed it. Whether you did that or not is up to you. Each and every repair is going to be different in one way or another than the ones before, but your skills will get better with each one you take on and finish.

Anytime you are using a spray can to prime or paint, remember these basic steps: Shake the can vigorously before spraying, spray a test pattern, apply a light or tack coat, and apply lighter coats with time for the primer or paint to flash (turn dull) before applying succeeding coats.

After shaking the paint can for about ten minutes, spray a test pattern on some sort of scrap. The last thing you want is for the paint to splatter on your workpiece.

1

Spray a light coat of the filler primer around all of the edges first. This minimizes overspray on the bulk of the workpiece.

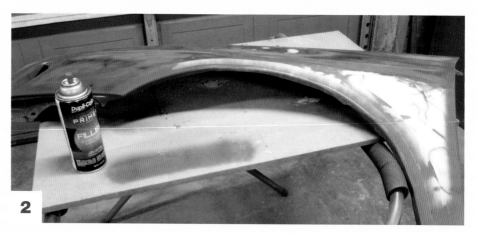

2

Apply a light coat of filler primer over the rest of the panel, using about a 50 percent overlap with each pass.

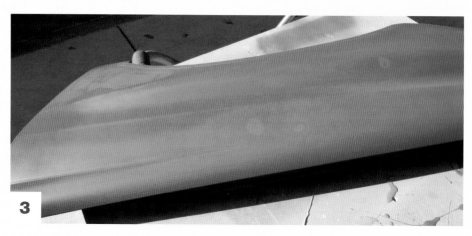

3

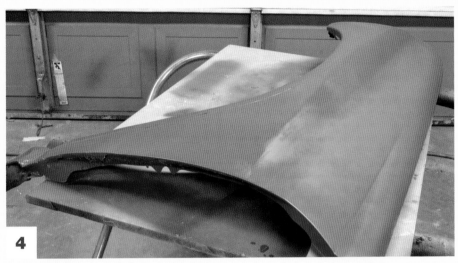

Apply the last coat of filler primer and then let it completely dry. After the last coat of filer primer dries, use a sanding block with 400-grit sandpaper and sand the entire panel once again.

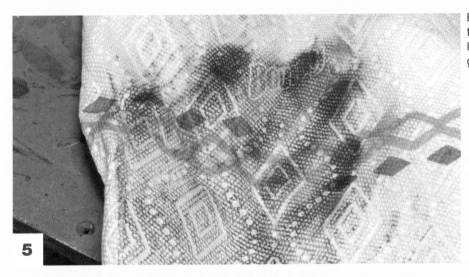

Blow away all sanding dust, then clean the surface one last time with wax and grease remover.

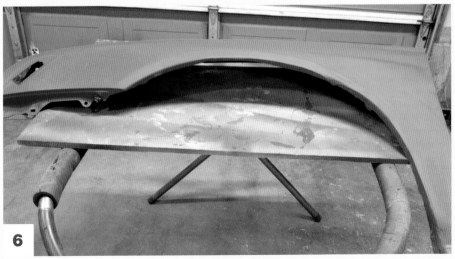

Now apply two coats of sandable primer, using the previous guidelines. Unless you see an issue, you don't have to sand anymore.

gauge of how you are progressing. If the primer is coming off evenly as you sand, the sheet metal is getting flatter. If the primer is coming off in local spots quicker, those spots are high. If the primer is not coming off as you sand, the area is low.

Paint

Spray paint is convenient because it is easy to use without expensive application equipment. Perhaps the biggest downfall is that it is difficult to get an exact color match. Some touchup paint is available in spray cans, but this is most likely in colors such as John Deere Green and Kubota Orange for tractors that do not have metallic, pearls, and candies in the paint mix.

Some automotive paint stores have the ability to put custom mixed paint into spray cans for you. This may be convenient, but one would suspect that the cost of this specialized service would outweigh any convenience, making it less expensive to simply

SPRAYING RATTLE CAN PAINT

Although spray cans pretty much operate the same way, regardless of the contents, paint is the final step so you want it to look the best it can. This finished application is not going to be sanded off, so imperfections are really going to show. No runs, drips, or other errors on this go 'round.

Verify the spray nozzle is clean by spraying a test pattern. Ensure that you are keeping the spray nozzle moving and at the same distance from the work surface. No pendulum arm swings.

1 Spray a test pattern on a piece of scrap cardboard, workbench, or wherever. Then apply a light tack coat to the entire surface.

purchase a pint of sprayable paint and rent a spray gun if necessary.

REMOVING MASKING MATERIAL

To many enthusiastic automobile painters, removing masking paper and tape to reveal a new paint job or quality spot paint repair is like opening birthday presents. It is always a pleasure to see a finished product, especially after viewing it in primer for any length of time. However, unlike the wrapping paper on presents, masking materials must be removed in a controlled manner to prevent finish damage.

To prevent paint flaking or peeling along the edge of masking tape strips, painters pull tape away from the newly painted body area (as opposed to straight up off the panel) and secure it back upon itself to create a sharp angle at the point tape leaves the surface. This sharp angle can cut extra-thin paint films so they don't cause flakes or cracks on the finish.

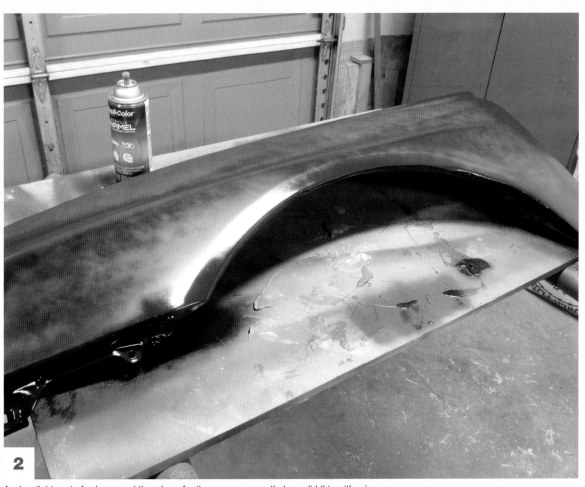

2 Apply a light coat of color around the edges, for the same reasons that you did this with primer.

Then apply a light coat of color to the rest of the panel, overlapping your spray pattern approximately 50 percent.

Let the first coat flash (turn dull) before applying a second or successive coats.

5

Let the painted piece dry for several hours prior to moving or touching it.

6

Invert the spray can and depress the cap to clear the spray nozzle.

Chapter 5
Prevention

CLEAN

Keeping your vehicle clean, inside and out, is one of the best ways to help prevent rust. Follow this up by applying a good wax product on a regular basis and your car will love you.

DEEP CLEAN

At least once a year, make it a point to remove everything that isn't part of your vehicle from the passenger compartment and trunk area. Use a high-powered vacuum to remove everything from underneath the seat. Empty out the trunk and vacuum it as well.

If any of these areas seem damp, use a wet-or-dry vacuum to pull out any moisture. Then do some investigating to determine the source of the moisture. If you simply spilled something, just remove it and don't do that again. However, if the moisture is coming from outside the vehicle, you may have some floor damage that requires repair.

WHEN TO WASH NEW PAINT FINISHES

Newer paints with hardener additives can generally be safely washed after one or two days, as long as mild automotive soap products are used and gentle washing efforts practiced. For uncatalyzed enamels, plenty of time (a few days or a week) should be allotted for paint solvents to evaporate or chemically react before newly sprayed car bodies are washed.

Who says the primer look is low maintenance? Even though this truck is painted in a semi-gloss form of black, to presumably resemble primer, these guys were quick to bring out the cleaning supplies to get rid of any road rash upon arrival at a major car event. Regardless of the quality of the bodywork or paint, any vehicle will look better if it is clean.

Your choice of car wash liquids, a large sponge (or car wash mitt), a bucket, and a hose are all you really need to improve the looks of your car in as few as thirty minutes or so. While automatic car washes are good (some better than others), doing a good, old-fashioned hand wash on your vehicle will give you a better idea of the condition of your vehicle's paint and body condition.

WASH

Car wash soap products can be found at auto parts stores and discount department stores. For the most part, almost any brand of car wash soap should be well suited for the finish on your vehicle. Just be sure to read the label for any warnings and to follow the mixing directions on labels of any product that you use.

The best way to prevent minute scratches or other blemishes on paint is to wash the vehicle in sections. Wash the dirtiest parts first, like the rocker panels, fender well lips, and lower front and rear end locations. Then, thoroughly rinse your soft cotton wash mitt and wash soap bucket. Mix up a new batch of wash soap to clean the vehicle sides. If their condition was relatively clean to start with, you can continue with that bucket of sudsy water to wash the hood, roof, and trunk areas.

This process rids your wash mitt and bucket of dirt and other scratch hazards, such as sand and road grit. If you were to wash your entire car with just one bucket of sudsy water, you increase the chances of your wash mitt picking up debris from the bucket where it will then be rubbed against the vehicle's lustrous finish. Likewise, anytime you notice that your wash mitt is dirty or if it should fall to the ground, always rinse it off with clear water before dipping it into the wash bucket. This helps to keep the wash water clean and free of debris.

To clean inside tight spaces, such as window molding edges and louvers, use a soft, natural hair floppy paintbrush. Do not use synthetic-bristled paintbrushes because they could cause minute scratches on paint surfaces. In addition, wrap a thick layer of heavy duct tape over the metal band on paintbrushes. This will help to guard against paint scratches or nicks as you vigorously agitate the paintbrush in tight spaces, possibly knocking the brush into painted body parts such as those around headlights and grilles.

PREVENTION

Other than vehicles that grow old sitting in a museum, all vehicles are going to be subjected to flying gravel or other road debris, accidental, or deliberate damage. While some of that damage will require professional help, minor rock chips or scratches can be repaired with a just a little bit of touchup paint.

If just a little bit of paint has been separated from the vehicle's body, but there is no other damage to the sheet metal, some touchup paint in the appropriate color and a small piece of 800-grit or finer sandpaper is all you need. Auto parts stores used to have a pretty good stock of touchup paint, but with the ever-increasing variety of colors, many have quit carrying touchup paint. If this is the case with your local auto parts store, you may need to purchase touchup paint from the dealer where you purchased your vehicle. While this might be slightly more expensive, it will be the best bet in getting the correct color. In either case, you may need to review the paint and options tag for your vehicle to find the correct color code.

Once you find the correct color, the rest is pretty easy. Most paint touchup bottles include a small brush, built right into the cap. This makes the paint easy to apply, but like all great paint jobs, there is some amount of prep work to be completed first. Just as primer and paint being sprayed from a spray gun, touchup paint will not properly adhere to wax, grease, or dirt. Saturate a paper towel with wax and grease remover and wipe off the area to be repaired. Then wipe the area dry with a clean paper towel. If you have a piece of 800-grit sandpaper or finer, use it to lightly scuff the surface where the touchup paint is to be applied. Avoid scuffing undamaged paint. Clean the area again, then brush on a light coat of touchup paint to the affected area and give it ample time to dry.

Chances are that this first brushed on coat will appear rough and uneven. Use the same extremely fine sandpaper that you used previously to sand down the new paint to a smooth surface. It is not your intent or desire to sand the touchup paint off, just make it smooth. When it is smooth, clean away any sanding residue, then apply another light coat of touchup paint, just as before. Let the touchup paint dry and repeat the process until the rock chip or scratch is no longer noticeable.

I was able to find this touchup paint for my 2008 Chevrolet Silverado along with several other common colors at a local discount department store. This particular label includes the name of the color (Victory Red) that is the same as my truck and the numbers below the name match the numbers from the truck's paint and options tag. Presumably, it is the correct touchup paint for my application.

1

Large or small, any area to be painted must be clean in order for the paint to adhere to the surface. Spray on some wax and grease remover, then wipe it off with a clean paper towel.

2

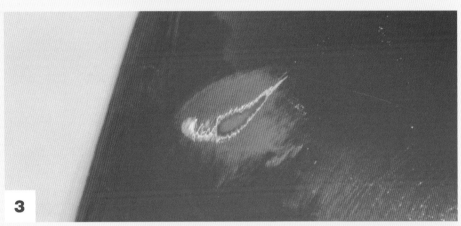

Use 800-grit or finer sandpaper to knock off the slick surface of the area so that the paint has some tooth to grip to. Notice that the area to be touched up has a slightly cloudy look to it after being scuffed up.

3

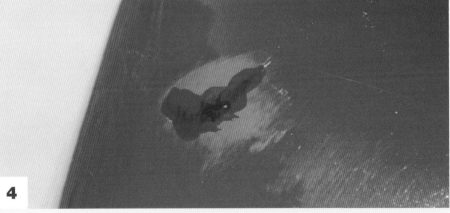

Make sure that you shake the bottle of touchup paint really well prior to each use. Using the brush in the cap of the touchup paint, apply a light coat to the affected area. Apply a light coat, so that it will dry quickly. Complete coverage may require several applications.

4

If there are any globs of paint after the touchup paint has dried, gently sand the surface again. Unlike a dent that may be filled with body filler, this type of repair is being filled with paint only.

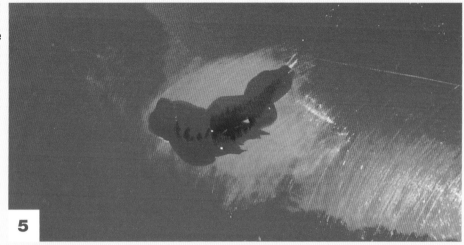

Allow plenty of time for the touchup paint to dry, then sand lightly again, and apply more touchup paint. Repeat the process until coverage is achieved and the surface finish is repaired.

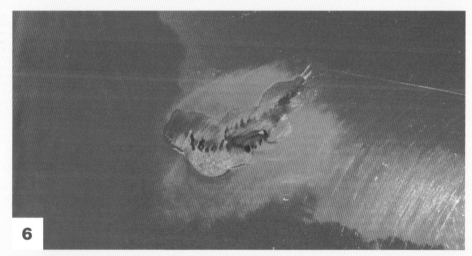

In this final photo, the repair is covered, but the paint doesn't match as well as I would like. However, it is not bad considering that the truck is red in color, is close to five years old, and sits outside all the time. The OEM paint has faded somewhat, so matching it exactly may be difficult at best. Whenever you are finished with your paint touchup, after the paint is thoroughly dry, wash the vehicle and apply a good coat of high quality wax.

WAX

One of the most confusing areas surrounding automotive finish maintenance for novice auto enthusiasts seems to focus on polish and wax products. Although both are designed as paint finish maintenance materials, each has its own separate purpose. Polishes clean paint finishes and remove accumulations of oxidation and other contaminants. On the other hand, wax does no cleaning or shining. It does, however, protect those paint finishes that have already been cleaned and polished.

Auto body paint and supply stores generally carry the largest selection of auto polishes and waxes, although many auto parts stores stock good assortments. Every polish should include a definitive label that explains what kind of paint finish it is designed for—for example, heavily oxidized, mildly oxidized, and new finish glaze. Those designed for heavy oxidation problems contain much coarser grit than those for new car finishes.

Along with descriptions of just which kind of paint finish particular polishes are designed for, labels will also note which products are intended for use with a buffer. Those with heavy concentrations of coarse grit are not recommended for machine use. Their polishing strength, combined with the power of a buffer, could cause large-scale paint burning problems.

Carnauba wax is perhaps the best product to use for protecting automobile paint finishes. Meguiar's, Eagle 1, and other cosmetic car-care product manufacturers offer auto enthusiasts an assortment of carnauba-based auto wax products. Other paint protection products are available that profess to work like wax, but contain different chemical bases, which you must clearly understand before applying them to your new paint job.

Several companies offer a wide variety of auto detailing products, but my personal favorite is Meguiar's. Shown are their paste wax (center), liquid wax (right), and Quik Detailer (left). The latter is great for quick use after washing your vehicle, presuming that it already has a good coat of wax on it. This is great for shining up your vehicle for a show or special date, without building up too much wax that can begin to look dull.

Some of these (typically, they have poly or polymer in the product name) are loaded with silicone materials. Although they may protect your car's finish for a long time, professional auto painters advise against their use because the silicone content is so high and saturating that any repainting that may be required in the future could be plagued by severe fisheye problems. In some cases, silicones have been known to penetrate paint finishes to eventually become embedded in sheetmetal panels.

If you find yourself in a quandary when it comes time to select a polish or wax product, seek advice from a knowledgeable auto body paint supplier. This person should be up to date on the latest product information from manufacturers and user satisfaction from professional painters and detailers in the field.

How Long Before Waxing?

Light coats of quality auto wax actually form protective seals on top of paint finishes. Even though they are quite thin and by no means permanent, these wax seals will prevent solvent evaporation. Should that occur, those vapors that need to exit paint would be trapped. Consequently, as confined vapors continue their evaporation activity and persistence in reaching the open atmosphere, minute amounts of pressure are built up, which eventually cause damage in the form of blisters to the new paint finish. So, instead of protecting a paint surface, waxing too soon after new paint applications can actually cause unexpected damage.

You should wait at least 90 to 120 days before waxing your freshly painted vehicle. During the summer or in any locations where the weather is warm and humidity is relatively low, 90 days should allow plenty of time for paint solvents to completely evaporate. When the weather is cooler or more humid, it takes longer for the solvent in the paint to evaporate, making longer waiting time before applying wax.

While you are waiting to apply a good coat of high quality wax to the exterior, you can do plenty of other things to make your once rusty car seem new again.

Cleaning the windows inside and out, vacuuming the carpet, and cleaning the upholstery will make any vehicle more enjoyable to drive. Polish the chrome, scrub the tires, then clean the engine compartment, and remove as much stuff as practical from the trunk. This once-damaged vehicle is looking good again. I knew you could do it …

LONG-TERM PAINT CARE

Although newer catalyzed paint products are much more durable and longer lasting than the materials used before them, you cannot expect their finish to shine forever without a minimal amount of routine maintenance. Basically, this entails washing, some polishing as needed, and periodic waxing.

Even though some paint products may be advertised as never requiring wax, many auto enthusiasts and professionals believe that good coats of wax not only help provide great paint longevity, but also make washing car bodies a lot easier. It almost seems like dirt and road debris float off waxed surfaces instead of having to be rubbed off.

Now that your commuter vehicle is back on the road after rust surgery, sooner or later, nicks or small chips will appear. Along with regular maintenance, you must also repair these minor paint problems as soon as possible. If not, exposed metal will oxidize and that corrosion will spread under paint to affect adjacent metal areas … again.

KEEPING YOUR CAR UNDERCOVER

Although you cannot keep your car under cover all the time, keeping it protected from the elements when not in use is certainly a good thing. A heated garage is ideal, but even a carport is better than nothing. If you are in a situation that makes a garage or carport unattainable, consider purchasing a car cover designed specifically for your vehicle. Shop around for one that meets your area's climatic needs before you buy. Purchasing the wrong type of cover can do more harm than good.

Sources

3M/Speedway Automotive
5320 W. Washington Street, Indianapolis, IN 46241
www.speedwayautoparts.com
317-243-6696

Masking tape, paper, and other products
Automotive Technology, Inc.
544 Mae Court, Fenton, MO 63026
www.automotivetechnology.com
800-875-8101, 636-343-8101

Paint booths, equipment, and supplies
Auto Parts Warehouse
16941 Keegan Avenue, Carson, CA 90746
www.autopartswarehouse.com
800-913-6119

Auto body parts
BASF/Automotive Finish
100 Park Ave., Florham Park, NJ 07932
www.basf.com
800-526-1072

Paint products
California Car Cover Company
9525 DeSoto Avenue, Chatsworth, CA 91311-5011
www.calcarcover.com
800-423-5525

Car covers
Campbell Hausfeld
100 Production Drive, Harrison, OH 45030
www.chpower.com
800-543-6400

Air compressors and pneumatic tools
Dupli-Color Products
101 Prospect Avenue NW500 Republic, Cleveland, OH 44115
www.duplicolor.com
800-247-3270

Paint products
Eastwood Company
263 Shoemaker Road, Pottstown, PA 19464
www.eastwoodcompany.com
800-343-9353

Automotive restoration tools, equipment, and supplies
Finesse Pinstriping, Inc.
www.finessepinstriping.com
800-228-1258

Striping tape
Hemmings Motor News
P.O. Box 100, Bennington, VT 05201
www.hemmings.com
800-227-4373

Classified ads for vehicles, products, and services
High Ridge NAPA
2707 High Ridge Boulevard, High Ridge, MO 63049
636-677-6400

Automotive parts, DuPont paint products
HTP America
180 Joey Drive, Elk Grove Village, Illinois 60007-1304
www.usaweld.com
800-872-9353, 847-357-0700

Welders, plasma cutters, tools, and accessories
Jerry's Auto Body, Inc.
1399 Church Street, Union, MO 63084
636-583-4757

Auto body repair
Licari Auto Body Supply, Inc.
2800 High Ridge Boulevard, High Ridge, MO 63049
636-677-1566

PPG paint products and supplies
Meguiar's
17991 Mitchell South, Irvine, CA 92614-6015
www.meguiars.com
800-347-5700

Car care products
Miller Electric Manufacturing Company
1635 W. Spencer Street, Appleton, WI 54912-1079
www.millerwelds.com
920-734-9821

Welders, plasma cutters, tools, and accessories
Mill Supply, Inc.
19801 Miles Rd, Cleveland, OH 44128
www.rustrepair.com
216-518-5072

Automotive rust repair panels
Mothers Polish Company
5456 Industrial Drive, Huntington Beach, CA 92649-1519
www.mothers.com
714-891-3364

Polishes, waxes, and cleaners
PPG Refinish Group
19699 Progress Drive, Strongsville, OH 44149
www.ppgrefinish.com
800-647-6050

Paint products
Sherwin-Williams Automotive Finishes
4440 Warrensville Center Road, Warrensville, OH 44128
www.sherwin-automotive.com
800-798-5872

Paint products
Stencils & Stripes Unlimited
1108 S. Crescent Avenue #38, Park Ridge, IL 60068
www.stencilsandstripes.com
847-692-6893

Reproduction paint stencils, stripes, and decals
Trim Parts
2175 Deer Field Road, Lebanon, OH 45036
www.trimparts.com, sales@trimparts.com
513-934-0815

GM Restoration Parts
Trim-Lok
6855 Hermosa Circle, Buena Park, CA 90620
www.trimlok.com
888-874-6565

Plastic and rubber trim and seals
Valspar Refinish
1101 South 3rd Street, Minneapolis, MN 55415
www.houseofkolor.com
800-444-2399

Paint products
Wyoming Technical Institute (Wyotech)
4373 N. 3rd Street, Laramie, WY 82072
www.wyomingtech.com
800-521-7158, 307-742-3776

Automotive technical school
Year One
P.O. Box 129, Tucker, GA 30085-0129
www.yearone.com
800-932-7663, 800-950-7663

Index